SpringerBriefs in Population Studies

Population Studies of Japan

The world population is expected to expand by 39.4% to 9.6 billion in 2060 (UN World Population Prospects, revised 2010). Meanwhile, Japan is expected to see its population contract by nearly one third to 86.7 million, and its proportion of the elderly (65 years of age and over) will account for no less than 39.9% (National Institute of Population and Social Security Research in Japan, Population Projections for Japan 2012). Japan has entered the post-demographic transitional phase and will be the fastest-shrinking country in the world, followed by former Eastern bloc nations, leading other Asian countries that are experiencing drastic changes.

A declining population that is rapidly aging impacts a country's economic growth, labor market, pensions, taxation, health care, and housing. The social structure and geographical distribution in the country will drastically change, and short-term as well as long-term solutions for economic and social consequences of this trend will be required.

This series aims to draw attention to Japan's entering the post-demographic transition phase and to present cutting-edge research in Japanese population studies. It will include compact monographs under the editorial supervision of the Population Association of Japan (PAJ).

The PAJ was established in 1948 and organizes researchers with a wide range of interests in population studies of Japan. The major fields are (1) population structure and aging; (2) migration, urbanization, and distribution; (3) fertility; (4) mortality and morbidity; (5) nuptiality, family, and households; (6) labor force and unemployment; (7) population projection and population policy (including family planning); and (8) historical demography. Since 1978, the PAJ has been publishing the academic journal *Jinkogaku Kenkyu* (The Journal of Population Studies), in which most of the articles are written in Japanese.

Thus, the scope of this series spans the entire field of population issues in Japan, impacts on socioeconomic change, and implications for policy measures. It includes population aging, fertility and family formation, household structures, population health, mortality, human geography and regional population, and comparative studies with other countries.

This series will be of great interest to a wide range of researchers in other countries confronting a post-demographic transition stage, demographers, population geographers, sociologists, economists, political scientists, health researchers, and practitioners across a broad spectrum of social sciences.

Ryo Oizumi

Population Dynamics Based on Individual Stochasticity

Springer

Ryo Oizumi
National Institute of Population and Social
Security Research
Tokyo, Japan

ISSN 2211-3215 ISSN 2211-3223 (electronic)
SpringerBriefs in Population Studies
ISSN 2198-2724 ISSN 2198-2732 (electronic)
Population Studies of Japan
ISBN 978-981-19-3547-3 ISBN 978-981-19-3548-0 (eBook)
https://doi.org/10.1007/978-981-19-3548-0

This Springer imprint is published by the registered company Springer Nature Singapore Pte Ltd.
The registered company address is: 152 Beach Road, #21-01/04 Gateway East, Singapore 189721, Singapore

Preface

At present, diversity is one of the most important issues worldwide. Diversity in race, gender, political ideology, and culture has created conflict and division throughout history. Tolerance for diversity is now regarded as the basis for stabilizing society. Diversity does not only explain the complex political or cultural components of human society: since 2020, the symptoms of COVID-19, a global pandemic, have also exhibited diversity. Some COVID patients may be asymptomatic, whereas others may become severely ill and die. These individual differences can be regarded as a form of diversity. We have long known that such individual differences are the driving forces in the evolution of organisms. Biologists have asserted that diversity within a population is necessary for adapting to environmental changes and infectious diseases. By referring to diversity as individual stochasticity, this book aims to incorporate it into mathematical demographic models. The author believes that the readers of this book will find it useful to broaden their perspective on demography.

Tokyo, Japan

Ryo Oizumi

Introduction: Stochasticity in Demography

Since the end of the nineteenth century and the onset of the twentieth century, scientific developments have provided us with revolutionary prospects for understanding the universe. There are three different analytical scales, namely macroscopic, microscopic, and mesoscopic, e.g., the theory of relativity, quantum theory, and chaos theory, respectively. As a representative macroscopic concept, the general theory of relativity explains problems on the universal scale. It has changed the landscape for time and space; both are not absolute but change with the physical environment of the observer. As a representative microscopic concept, quantum theory describes the dynamics of elementary particles. This theory postulates that the law of nature incorporates probability, which was difficult for people to accept at the time it was proposed. The mesoscopic scale is an ordinary scale that is observed in daily life. Chaos theory, statistical physics, fluid mechanics, and demography are categorized under this scale. In particular, chaos theory shows that uncertainty occurs in deterministic systems, and this idea was also unacceptable previously. Among the three aforementioned scales, the mesoscopic scale is the most complicated one because it intermixes with both of the other two scales. Moreover, it includes not only deterministic systems, but also stochastic systems, such as the variable environment. Therefore, systems on this scale are often called complex systems.

These topics are thought to be involved only in physical phenomena and not in demography. However, the complicated activities and population dynamics of humans occur on the mesoscopic scale, where the three aforementioned concepts intersect. Each behavior, as a form of microscopic dynamics, yields macroscopic population dynamics as physical phenomena. Thus, demography can also be regarded as a field of physics.

For example, internal or domestic migration occurs for various reasons, such as job transfer, school attendance, and marriage. Although these events are not random, we evaluate them using statistics instead of identifying their causes. This approach has the same context as statistical physics. Each classical particle follows Newtonian mechanics; however, the heat bath dynamics obey statistics such that each particle behaves randomly. Meanwhile, human beings and other species have individual differences that may have a stochastic effect on their population dynamics.

For example, individuals with a healthy immune system survived the Great Plague and other epidemics. We know that such events often occurred in human history. The heterogeneity of individuals brings innovation to human society, protects the population from extinction, and promotes evolution. Interpreting heterogeneity as individual stochasticity can provide a new perspective on demography.

This book aims to reconstruct mathematical demography on the basis of individual stochasticity. To this end, we use our knowledge of stochastic analysis. However, this is not a mathematical treatise; the mathematical justification is limited to the minimum required to be applied to this field. There are many excellent mathematical works on stochastic analysis, and the mathematical details are presented therein. Because our theory focuses on the description of biological and demographic phenomena, we often use techniques that are not always justified mathematically. Abstract mathematical theory is useful for explaining a wide range of demographic and biological events. Meanwhile, for the benefit of readers who are unfamiliar with such mathematical concepts, examples that are analytically solvable are provided in each chapter. Although they are not necessarily for human subjects, they will enable the reader to understand the theory developed in this book. Finally, we expect this book to enable readers to broaden their perspective on population dynamics.

Tokyo, Japan

Ryo Oizumi
ooizumi-ryou@ipss.go.jp

Contents

Chapter 1
Deterministic and Stochastic Population Models

Newtonian mechanics and the theory of relativity, which describe planetary motion accurately, have enabled humans to land on the moon. There is a strict rule that mathematically connects gravity and motion. However, demographers have not identified such a connection in population dynamics thus far (or, it may not exist). Accordingly, we still do not have any theory that can predict the exact figures for future populations. Nevertheless, researchers have used mathematics and have systematized the theory of mathematical demography. One of the reasons is to not only identify the strict rule in demographic phenomena, but also to establish a universal index that determines the eventual state of a population. The Malthus equation is the simplest and most basic mathematical model in this field. This chapter discusses the difference between deterministic and stochastic demographic models through the addition of noise.

1.1 Malthus Equation and Stochastic Malthus Equations

If N_t is the population size at time t, the population growth rate is proportional to the population size; hence, the equation consists of a proportional constant r and the population size itself

$$\frac{dN_t}{dt} = rN_t. \tag{1.1}$$

If the *crude fertility rate* (CFD) is greater than the *crude death rate* (CDR), r becomes positive and the population increases. Conversely, when the CFD is less than the CDR, r is negative and the population decreases. This equation is well known and is called the Malthus equation. Further, r is also called the Malthusian parameter. Therefore, the Malthusian parameter can be regarded as a universal index because it determines

R. Oizumi, *Population Dynamics Based on Individual Stochasticity*,
Population Studies of Japan,
https://doi.org/10.1007/978-981-19-3548-0_1

whether the population increases in this model. The solution of the Malthus equation is the exponential growth

$$N_t = N_0 \exp\{rt\}.$$

We know that the Malthusian parameter is not always constant. In the real world, population growth is exposed to various uncertainty factors, even if it increases monotonically. This is not the same as the rule of planetary motion; however, it is a fact that population growth is proportional to the number of mothers. If we incorporate uncertainty in the population dynamics, what is going to occur in the long term? Such problems are treated using stochastic differential equations (SDEs). The Malthus equation can be formulated as an SDE as follows:

$$dN_t = rN_t dt + \underbrace{\sigma_0 N_t dB_t}_{\text{noise term}} \quad N_{t=0} = N_0, \tag{1.2}$$

where B_t denotes Brownian motion. The second term on the right-hand side represents the noise of the Malthusian parameter. Mathematical justification of the noise requires specialized knowledge about functional analysis as well as considerable preparation; however, these aspects are devolved to the referencesIkeda and Watanabe (1989); Karatzas and Shreve (1991); Øksendal (2003). As this book focuses on the mathematical knowledge of stochastic analysis as a demographic tool, here, we only enumerate the properties of stochasticity. The reason for the use of infinitesimal differences on the left-hand side of Eq. (1.2) is that N_t is generally indifferentiable in the ordinary sense. The Brownian motion is a process satisfying the normal distribution $\mathrm{N}\left(m, \sigma^2\right)$ with expected value m and variance σ^2 such that

$$B_t - B_s \sim N(0, t-s) \quad s < t, \quad B_0 = 0 \quad \text{(for probability 1)}. \tag{1.3}$$

In this stage, the equation is still incomplete because there is no definition of the solution of the SDE. The solutions of ordinary differential equations are defined by well-known integrals such as the Lebesgue integral or the Riemann–Stieltjes integral. However, for SDEs, we must consider the integral for the random noise as follows:

$$N_t = N_0 + r\int_0^t ds\, N_s + \sigma_0 \int_0^t N_s dB_s.$$

The integral of the second term on the right-hand side is called the stochastic integral Karatzas and Shreve (1991), and it is defined by the Riemann–Stieltjes integral

$$\int_0^t N_s dB_s := \lim_{n\uparrow\infty} \sum_{i=0}^{n-1} N_{\tau_i}\left(B_{t_{i+1}} - B_{t_i}\right) \quad \tau_i \in \left[t_i, t_{i+1}\right] \quad 0 = t_0 \le t_1 \le \cdots \le t_n = t.$$

As the Brownian motion is not a function of bounded variation, this Riemann sum does not converge in its original sense. It is regarded as convergence in the sense of the second-order integral such that

$$\lim_{n\uparrow\infty}\sum_{i=0}^{n-1}\mathbb{E}\left[N_{\tau_i}^2\left(B_{t_{i+1}}-B_{t_i}\right)^2\right]=\lim_{n\uparrow\infty}\sum_{i=0}^{n-1}\mathbb{E}\left[N_{\tau_i}^2\left(t_{i+1}-t_i\right)\right]=\mathbb{E}\left[\int_0^t N_s^2 ds\right]<\infty,$$

where $\mathbb{E}[\cdot]$ represents an expectation. The right-hand side is the usual Riemann–Stieltjes integral, and it is derived from the variance of the Brownian motion.

The most important aspect of this application is how to choose τ_i. The Riemann–Stieltjes integral does not depend on the choice of τ_i, whereas the stochastic integral has different results depending on τ_i. For example, if $\tau_i = t_i$, the solution of Eq. (1.2) becomes

$$N_t = N_0 \exp\left\{\left(r-\frac{\sigma_0^2}{2}\right)t+\sigma_0 B_t\right\}. \tag{1.4}$$

The derivation of the function above is presented later.[1] The well-defined *ordinary differential equation* (ODE) gives a unique solution depending on the initial state; however, the SDE gives a unique probability density function (PDF) for each initial state. Therefore, each solution, which is called a sample path, follows the probability distribution. The Brownian motion diffuses at the $t^{\frac{1}{2}}$ order as its property is such that the deterministic term of the exponent dominates each sample path Øksendal (2003). In other words, if $r < \frac{\sigma^2}{2}$, all sample paths converge to zero, and the population approaches extinction with a probability of one. As the Brownian motion follows the normal distribution, the expectation of the population corresponds to the solution of the Malthus equation

$$\mathbb{E}_{N_0}\left[N_t\right]=\int_{-\infty}^{\infty} dx\frac{N_0}{\sqrt{2\pi t}}\exp\left\{\left(r-\frac{\sigma_0^2}{2}\right)t+\sigma_0 x-\frac{x^2}{t}\right\}=N_0\exp\{rt\}.$$

This result seems to contradict the solution above; however, it often occurs in the computation of expectation. This phenomenon is due to sample paths that have few probabilities but generate a large population size. Meanwhile, if we choose τ_i as the central point of these intervals,

$$\tau_i=\frac{t_{i+1}+t_i}{2},$$

each sample path is provided by the following equation:

$$N_t = N_0 \exp\{rt+\sigma_0 B_t.\} \tag{1.5}$$

The solution above shows that random noise does not affect the extinction of the population. Thus, this system follows the conventional Malthusian parameter. The definitions of the stochastic integral are mathematically correct. The integral defined

---

[1] This function is known as a geometric Brownian motion.

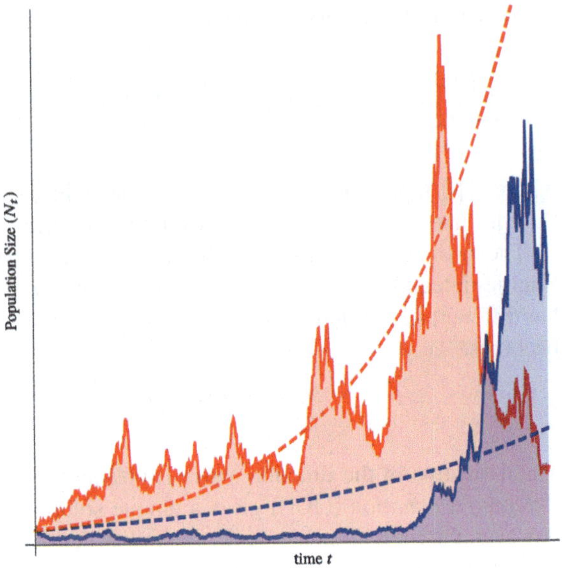

Fig. 1.1 Deterministic and stochastic population growth: This figure shows the sample paths of two different SDEs and their average trajectories. The blue and red solid lines represent the sample paths of the Ito and Stratonovich types of SDEs. The dashed lines represent their average trajectories

by the left endpoint $\tau_i = t_i$ is called Ito's integral, whereas the other one, $\tau_i = \frac{t_{i+1}+t_i}{2}$, is called Stratonovich's integral. There is a controversy as to which definition is more accurate in terms of representing the variable environment, which is yet to be resolved. Each stochastic integral has its advantages. Ito's integral does not include information about the future; hence, it is applied to the prediction of stock values and the approximation of Markovian processes based on discrete-time models such as matrix models. Meanwhile, because the Stratonovichs integral has the same mathematical properties as the normal integral, it can be naturally incorporated into the extension of dynamical systems. To the best of our knowledge, biologists, economists, and demographers advocate the former, whereas physicists tend to use the latter. Thus, by the definition of stochastic integrals, we have various SDEs that are statistically different. Hence, we should carefully observe the uncertainties involved in the phenomenon of interest (Fig. 1.1).

1.2 Basic Properties for SDE Models

The previous section showed that the SDE has several versions depending on the definition of the stochastic integral. Ito's integral and Stratonovich's integral give solutions that are statistically different from each other. This section discusses why such differences occur by using the properties of SDEs. Consider a rewritten form of the stochastic Malthus equation defined by Ito's integral

$$\frac{dN_t}{N_t} = rdt + \sigma_0 dB_t,\ N_{t=0} = N_0. \tag{1.6}$$

Let us integrate both sides. Hence, we have

$$\int_0^t \frac{dN_s}{N_s} = rt + \sigma_0 B_t. \tag{1.7}$$

Usually, the left-hand side becomes a logarithmic function; however, this is not correct in the case of the SDE. For instance, let us consider a derivative of the following logarithmic function for the random variable N_t:

$$d\ln\{N_t\} = \frac{dN_t}{N_t} - \frac{1}{2}\frac{dN_t^2}{N_t^2}. \tag{1.8}$$

In ordinary differentiation, we can ignore the second term on the right-hand side, whereas it is not true for the SDE because substituting Eq. (1.2) into the equation above gives the following equation:

$$d\ln\{N_t\} = rdt + \sigma_0 dB_t - \frac{\sigma_0^2}{2}dt.$$

Then, we use the property of infinitesimal difference in the Brownian motion as follows:

$$dt^2 = 0,\ dtdB_t \approx dt^{\frac{3}{2}} = 0,\ dB_t^2 = dt, \tag{1.9}$$

where the third equation is derived from the variance of the Brownian motion[2] Eq. (1.3). Thus, a function for random variables generates the first-order difference from the second-order term. This manipulation and the formula are called "stochastic differentiation" and "Ito's formula (or lemma)," respectively. Putting all the pieces together, we get

$$\int_0^t \frac{dN_s}{N_s} = \int_0^t d\ln\{N_s\} + \frac{\sigma_0^2}{2}\int_0^t ds = rt + \sigma_0 B_t. \tag{1.10}$$

Therefore, the solution of the Ito-type stochastic Malthus equation becomes

$$N_t = N_0 \exp\left\{\left(r - \frac{\sigma_0^2}{2}\right)t + \sigma_0 B_t\right\}.$$

In Ito's integral, Stratonovich's integral is computed by

---

[2] This relationship is called Lèvy's formula.

$$
\begin{aligned}
\sigma_0 N_{t_i}\left(B_{t_{i+1}}-B_{t_i}\right) &= \sigma_0\left(N_{\left(\frac{t_{i+1}+t_i}{2}-\frac{t_{i+1}-t_i}{2}\right)}\right)\left(B_{t_{i+1}}-B_{t_i}\right) \\
&= \sigma_0\left(N_{\frac{t_{i+1}+t_i}{2}}-\left(N_{\frac{t_{i+1}+t_i}{2}}-N_{\left(\frac{t_{i+1}+t_i}{2}-\frac{t_{i+1}-t_i}{2}\right)}\right)\right)\left(B_{t_{i+1}}-B_{t_i}\right) \\
&= \sigma_0\left(N_{\frac{t_{i+1}+t_i}{2}}-\frac{r}{2}N_{\frac{t_{i+1}+t_i}{2}}\left(t_{i+1}-t_i\right)-\frac{\sigma_0}{2}N_{\frac{t_{i+1}+t_i}{2}}\left(B_{t_{i+1}}-B_{t_i}\right)\right) \\
&\quad\times\left(B_{t_{i+1}}-B_{t_i}\right) \\
&= \sigma_0 N_{\frac{t_{i+1}+t_i}{2}}\left(B_{t_{i+1}}-B_{t_i}\right)-\frac{\sigma_0^2}{2}N_{\frac{t_{i+1}+t_i}{2}}\left(t_{i+1}-t_i\right).
\end{aligned}
$$

Using a symbol $\int_0^t N_s \circ dB_s$ for Stratonovich's integral, the two stochastic integrals have the following relationship:

$$
\sigma_0 \int_0^t N_s dB_s = \sigma_0 \int_0^t N_s \circ dB_s - \frac{\sigma_0^2}{2}\int_0^t ds\, N_s.
$$

The equation above shows that extinction by stochasticity does not occur in the case of Stratonovich's integral because the deviation from Ito's integral as the second term on the right-hand side cancels out the attenuation due to noise. We have illustrated several essential properties of SDEs via the stochastic Malthus equation. These properties should be generalized and formulated to apply SDE models to various demographic problems. Hence, let us consider a general Ito-type SDE such that each ingredient of the d-dimensional random variable X_t follows

$$
dX_t^j = g_j(t, X_t)\,dt + \sum_{k=1}^{n} \sigma_{jk}(t, X_t)\,dB_t^k, \quad X_0^j = x^j, \quad j = 1, 2, \cdots, d. \tag{1.11}
$$

Then, Ito's formula for a multi-variable function concerning $f(t, X_t)$ is generalized in the following stochastic differential form:

$$
\begin{aligned}
df(t, X_t) &= \frac{\partial}{\partial t} f(t, X_t)\,dt + \sum_{j=1}^{d} \frac{\partial}{\partial x^j} f(t, X_t)\,dX_t^j \\
&\quad + \frac{1}{2}\sum_{j=1}^{d}\sum_{j'=1}^{d} \frac{\partial^2}{\partial x^j \partial x^{j'}} f(t, X_t)\,dX_t^j dX_t^{j'} \\
&= \frac{\partial}{\partial t} f(t, X_t)\,dt + \sum_{j=1}^{d} g_j(t, X_t)\frac{\partial}{\partial x^j} f(t, X_t)\,dt \\
&\quad + \frac{1}{2}\sum_{j=1}^{d}\sum_{j'=1}^{d}\sum_{k=1}^{n} \sigma_{jk}(t, X_t)\,\sigma_{j'k}(t, X_t)\frac{\partial^2}{\partial x^j \partial x^{j'}} f(t, X_t)\,dt \\
&\quad + \sum_{j=1}^{d}\sum_{k=1}^{n} \sigma_{jk}(t, X_t)\frac{\partial}{\partial x^j} f(t, X_t)\,dB_t^k,
\end{aligned} \tag{1.12}
$$

where the elements of n-dimensional Brownian motion are independent of each other such that

$$dB_t^k dB_t^k = \delta_{kk'} dt \quad \delta_{kk'} : \text{Kronecker'sdelta}. \tag{1.13}$$

Moreover, the transformation of Ito's integral to Stratonovich's integral becomes

$$\begin{aligned}
\sum_{k=1}^{n} \sigma_{jk}\left(t_i, X_{t_i}\right)\left(B_{t_{i+1}}^k - B_{t_i}^k\right) &= \sum_{k=1}^{n} \sigma_{jk}\left(\frac{t_{i+1}+t_i}{2} - \frac{t_{i+1}-t_i}{2}, X_{\frac{t_{i+1}+t_i}{2} - \frac{t_{i+1}-t_i}{2}}\right) \\
&\quad \times \left(B_{t_{i+1}}^k - B_{t_i}^k\right) \\
&= \sum_{k=1}^{n} \sigma_{jk}\left(\frac{t_{i+1}+t_i}{2}, X_{\frac{t_{i+1}+t_i}{2}}\right)\left(B_{t_{i+1}}^k - B_{t_i}^k\right) \\
&\quad - \frac{1}{2}\sum_{j'=1}^{d}\sum_{k=1}^{n} \sigma_{jk}\left(\frac{t_{i+1}+t_i}{2}, X_{\frac{t_{i+1}+t_i}{2}}\right) \\
&\quad \times \frac{\partial}{\partial x^{j'}}\sigma_{j'k}\left(\frac{t_{i+1}+t_i}{2}, X_{\frac{t_{i+1}+t_i}{2}}\right)\left(t_{i+1}-t_i\right).
\end{aligned}$$

Therefore, we get

$$\begin{aligned}
\sum_{k=1}^{n}\int_0^t \sigma_{jk}(s, X_s)\, dB_s^k &= \sum_{k=1}^{n}\int_0^t \sigma_{jk}(s, X_s) \circ dB_s^k \\
&\quad - \frac{1}{2}\sum_{j'=1}^{d}\sum_{k=1}^{n}\int_0^t ds\, \sigma_{j'k}(s, X_s)\frac{\partial}{\partial x^{j'}}\sigma_{jk}(s, X_s).
\end{aligned} \tag{1.14}$$

The relationship above is called the Stratonovich transform. Thus, typical SDEs are formulated by adding the infinitesimal difference of Brownian motion that is independent and identically distributed (i.i.d) to ODEs. However, by the interpretation of random noise, we have various stochastic integrals with different solutions. These observations imply that the correct interpretation of the uncertainty in nature is yet to be established.

1.3 SDEs and Partial Differential Equations

In contrast to ODEs, SDEs do not have a unique solution for each initial condition. In other words, SDEs have numerous sample paths following a different distribution for each initial state. However, each distribution must be unique. If this is not true, SDEs would have no mathematical justification. Then, these distributions should be derived from the statistics of a path generated by the SDEs. This section treats the *partial differential equation* (PDE) derived from the property of the SDE that should be satisfied by the PDF. Let us consider a PDF $Q\,(s, x \to t, y)$ that represents the flow from x at time s to y at time t $(s < t)$. If this function follows the rule

$$Q\,(s, x \to t, y) = \int d\xi\ Q\,(s, x \to \tau, \xi)\, Q\,(\tau, \xi \to t, y) \quad s < \tau < t, \tag{1.15}$$

the flow belongs to a Markovian process, which refers to the memoryless property of a stochastic process. Then, Eq. (1.15) is called the Chapman–Kolmogorov equation, and it represents the summation over all PDFs of a path through ξ at time τ. Let us consider the integration for the neighborhood of y in Eq. (1.15) for a sufficiently short time Δt, and it is expanded into a Taylor series as follows:

$$\begin{aligned} Q\,(s, x \to t + \Delta t, y) &= \int d\eta\ Q\,(s, x \to t, y - \eta)\, Q\,(t, y - \eta \to t + \Delta t, y) \\ &= \int d\eta\ \exp\{-\eta \cdot \nabla\}\,(Q\,(s, x \to t, y)\, Q\,(t, y \to t \\ &\quad + \Delta t, y + \eta)) \\ &= \sum_{m=0}^{\infty} \frac{(-1)^m}{m!} \int d\eta \left(\sum_{j=1}^{d} \eta_j \frac{\partial}{\partial y^j} \right)^m (Q\,(s, x \to t, y) \\ &\quad \times Q\,(t, y \to t + \Delta t, y + \eta))\,, \end{aligned} \tag{1.16}$$

where an operator generating the Taylor series gives a shift of variables for a multi-variable function, such as

$$f\,(x - \eta) = \exp\{-\eta \cdot \nabla\}\, f\,(x) := \sum_{m=0}^{\infty} \frac{(-1)^m}{m!} \left(\sum_{j=1}^{d} \eta_j \frac{\partial}{\partial x^j} \right)^m f\,(x)\,.$$

Because $Q\,(\cdot)$ constitutes the probability measure, it should satisfy

$$\int dy\ Q\,(s, x \to t + \Delta t, y) = 1,$$

and the zero-order term $m = 0$ becomes $Q\,(s, x \to t, y)$. To eliminate complications, let us omit the initial state (s, x) in the function Q as follows:

$$Q\,(t, y) \equiv Q\,(s, x \to t, y)\,.$$

Furthermore, we exchange the orders of the integral and the partial differential on the right-hand side of Eq. (1.16)

$$\int d\eta \; \exp\{-\eta \cdot \nabla\} \, Q(s, y \to t, y + \eta) \, Q(t, y)$$
$$= \sum_{m=1}^{\infty} \frac{(-1)^m}{m!} \sum_{j_1, \cdots, j_m = 1}^{d} \frac{\partial^m}{\partial y^{j_1} \cdots \partial y^{j_m}} \underbrace{\int d\eta^1 \cdots d\eta^d \, \eta^{j_1} \cdots \eta^{j_m} \, Q(s, y \to t, y + \eta) \, Q(t, y)}_{C_m^{j_1, \cdots, j_m}(t, y) \Delta t} .$$

Because $Q(s, y \to t, y + \eta)$ represents the transition probability from y to $y + \eta$ at time t, and η is interpreted as an infinitesimal difference for each state, the coefficient $C_m^{j_1, \cdots, j_m}(t, y)$ implies the m-th order of the correlation among these states

$$C_m^{j_1, \cdots, j_m}(t, y) := \lim_{\Delta t \downarrow 0} \mathbb{E}_{t, y} \left[\frac{\left(X_{t+\Delta t}^{j_1} - X_t^{j_1}\right) \cdots \left(X_{t+\Delta t}^{j_m} - X_t^{j_m}\right)}{\Delta t} \right] . \qquad (1.17)$$

This coefficient is referred to as the m-th order *Kramers–Moyal coefficient*. From these pieces, using

$$\lim_{\Delta t \downarrow 0} \frac{Q(s, x \to t + \Delta t, y) - Q(s, x \to t, y)}{\Delta t} = \frac{\partial}{\partial t} Q(s, x \to t, y) ,$$

the Chapman–Kolmogorov equation is converted into a PDE

$$\frac{\partial}{\partial t} Q(t, y) = \sum_{m=0}^{\infty} \frac{(-1)^m}{m!} \sum_{j_1, \cdots, j_m = 1}^{d} \frac{\partial^m}{\partial y^{j_1} \cdots \partial y^{j_m}} \left(C_m^{j_1, \cdots, j_m}(t, y) \, Q(t, y) \right)$$
$$Q(s, y) = \delta^d (x - y) . \qquad (1.18)$$

As a state transition cannot stay at two different points at the same time, the probability density $Q(s, x \to s, y)$ is not weighted except at the point $x - y$, i.e.,

$$Q(s, x \to s, y) = \delta^d (x - y) ,$$

where $\delta^d (x - y)$ denotes the d-dimensional Dirac's delta as follows:

$$\delta^d (x - y) := \prod_{j=1}^{d} \delta \left(x^j - y^j \right) .$$

Equation (1.18) is called the Kramers–Moyal equation Goel et al. (1974). The expansion of Eq. (1.18) is called the Kramers–Moyal expansion Kramers (1940); Moyal (1949). With the PDF, the zero-order coefficient is always zero: $C_0(t, y) = 0$. However, it is not necessary to maintain this condition in general, and it plays a significant role in demography. The first order and second order of Kramers–Moyal coefficients represent the mean and covariance of the stochastic differential, respectively:

$$C_1^j(t, y) = \lim_{\Delta t \downarrow 0} \mathbb{E}\left[\frac{\left(X_{t+\Delta t}^j - X_t^j\right)}{\Delta t}\right],$$

$$C_2^{jj'}(t, y) = \lim_{\Delta t \downarrow 0} \mathbb{E}\left[\frac{\left(X_{t+\Delta t}^j - X_t^j\right)\left(X_{t+\Delta t}^{j'} - X_t^{j'}\right)}{\Delta t}\right].$$

Let us recall the property of the SDE concerning Ito's formula. If $X_t^j = y^j$ in Eq. (1.11), the first-order Kramers–Moyal coefficient becomes

$$C_1^j(t, y) = g_j(t, y) \tag{1.19}$$

because $\mathbb{E}[dB_t] = \mathbb{E}\left[B_{t+\Delta t} - B_t\right] = 0$. As for its covariance, it corresponds to the coefficient of the second derivative in Ito's formula, i.e.

$$C_2^{jj'}(t, y) = S_{jj'}(t, y) := \sum_{k=1}^{n} \sigma_{jk}(t, y)\, \sigma_{j'k}(t, y). \tag{1.20}$$

Recalling the properties of the SDEs given by Eqs. (1.9) and (1.13), the order of the coefficients of the master equation generated by Eq. (1.11) is not higher two . Therefore, the PDF of the SDE follows

$$\frac{\partial}{\partial t} Q(t, y) = -\sum_j \frac{\partial}{\partial y^j}\left(g_j(t, y)\, Q(t, y)\right) + \frac{1}{2}\sum_{j=1}^{d}\sum_{j'=1}^{d} \frac{\partial^2}{\partial y^j \partial y^{j'}}\left(S_{jj'}(t, y)\, Q(t, y)\right), \tag{1.21}$$

$$Q(s, y) = \delta^d(x - y). \tag{1.22}$$

This equation is the well-known Fokker–Planck equation or Kolmogorov forward equation. It gives the distribution that the solution of the Ito-type SDE obeys for each time t. Broadly speaking, the uniqueness of the distribution justifies the uniqueness of the SDE solutions. Overall, ODEs such as $\frac{dN_t}{dt}$ in Eq. (1.1) follow the right-hand side rule, whereas SDEs (e.g., dX_t^j in Eq. (1.11)) are interpreted by following a Gaussian distribution with mean $g_j(t, X_t)$ and covariance $S_{jj'}(t, X_t)$.

The derivation method of Eq. (1.21) enables us to set (s, x) to the variable instead of (t, y). By applying the same procedure as that for the derivation of the Fokker–Planck equation to $Q(s - \Delta s, x \to t, y)$ in Eq. (1.15), the Chapman–Kolmogorov equation is rewritten as

$$Q(s-\Delta s, x \to t, y) = \int d\xi \; Q(s-\Delta s, x \to s, \xi)\, Q(s, \xi \to t, y). \quad (1.23)$$

By expanding $Q(s, \xi \to t, y)$ on the right-hand side into a Taylor series for ξ in the neighborhood of x, the backward Kramers–Moyal equation is derived, such that

$$\begin{aligned} -\frac{\partial}{\partial s} Q(s,x) &= \sum_{m=0}^{\infty} \frac{1}{m!} \sum_{j_1,\cdots,j_m=1}^{d} C_m^{j_1,\cdots,j_m}(s,x) \frac{\partial^m}{\partial x^{j_1}\cdots\partial x^{j_m}} Q(s,x) \\ Q(t,x) &= \delta^d(x-y). \end{aligned} \quad (1.24)$$

From the SDE property used in the derivation of the Fokker–Planck equation, we get another equation concerning the initial state, which is called the Kolmogorov backward equation

$$\begin{aligned} -\frac{\partial}{\partial s} Q(s,x) &= \sum_j g_j(s,x) \frac{\partial}{\partial x^j} Q(s,x) + \frac{1}{2} \sum_{j=1}^{d} \sum_{j'=1}^{d} S_{jj'}(s,x) \frac{\partial^2}{\partial x^j \partial x^{j'}} Q(s,x) \\ Q(t,x) &= \delta^d(x-y). \end{aligned} \quad (1.25)$$

Basically, the Fokker–Planck equation gives the future distribution of X_t with the initial condition (s, x), whereas the backward equation gives that of the past with the terminal condition (t, y). As for the stochastic Malthus equation, i.e., Eq. (1.2), the following Fokker–Planck equation gives its PDF:

$$\begin{aligned} \frac{\partial}{\partial t} Q(t,N) &= -\frac{\partial}{\partial N} \left(rNQ(t,N) \right) + \frac{1}{2} \frac{\partial^2}{\partial N^2} \left(\sigma_0^2 N^2 Q(t,N) \right) \\ Q(0,N) &= \delta(N_0 - N). \end{aligned}$$

This equation is explicitly solvable, and its solution is known as the logarithmic normal distribution such that

$$Q(t,N) = \frac{N_0}{N\sqrt{2\pi\sigma_0^2 t}} \exp\left\{ \frac{-\left(\ln\left\{\frac{N}{N_0}\right\} - \left(r - \frac{\sigma_0^2}{2} \right) t \right)^2}{2\sigma_0^2 t} \right\}$$

Although it has already been mentioned in the first section of this chapter, we can observe that the distribution focuses on the population density being equal to zero if $r < \frac{\sigma_0^2}{2}$.

Overall, the SDE and the parabolic-type PDE are related to each other. The second-order term that the SDE generates is called the "diffusion term," and it incorporates probability theory into the PDE. In this chapter, the stochastic Malthus equation is used as an example to explain SDEs and their analysis. As its solution is a sample that follows statistics, the probability distribution determines the population projection statistically. Therefore, the SDE can be interpreted to generate individuals that constitute its ensemble following the PDE. In the next chapter, we introduce a structured population model based on this perspective.

References

Ikeda, N., & Watanabe, S. (1989). *Stochastic differential equations and diffusion processes* (2nd ed.). North-Holland: Elsevier.

Karatzas, I., & Shreve, S. (1991). *Brownian motion and stochastic calculus* (Vol. 113). Berlin: Springer Verlag.

Øksendal, B. (2003). *Stochastic differential equations: an introduction with applications*. Berlin: Springer Verlag.

Goel, N., & Richter-Dyn, N. (1974). *Stochastic models in biology*. London: Academic Press.

Kramers, H. A. (1940). Brownian motion in a field of force and the diffusion model of chemical reactions. *Physica, 7*(4), 284–304.

Moyal, J. (1949). Stochastic processes and statistical physics. *Journal of the Royal Statistical Society. Series B (Methodological), 11*(2), 150–210.

Chapter 2
Linear Structured Population Based on SDE

Organisms constituting a species also include various internal heterogeneities. These include observable entities such as phenotypes to abstract states such as assets. They are not uniform for each cohort and are closely associated with critical demographic events such as reproduction and death. For instance, larger stags will have a higher degree of success in mating with females, and high-income earners will have higher survival rates because they can access advanced medical care. Although heterogeneity often inhibits the development of antibiotics[1] and promotes social inequality, it also protects populations from extinction and triggers the evolution of species. Incorporating such individual differences in demographic theory can enable us to extend the theory to deal with a broad range of phenomena.

This chapter treats population dynamics consisting of physiological ingredients that yield individual heterogeneity. Each ingredient is represented by an SDE model; thus, heterogeneity is referred to as individual stochasticity in this book. This framework is one of the most straightforward definitions of SDE. Indeed, SDE has various versions that we have not introduced and which do not always follow the normal distribution.

There are mathematical methods in demography including the Coale–Preston method Preston et al. (1980), Multi-State Life Tables Rogers and Ledent (1976), and so on. In relation to those methods, the model presented here corresponds to an infinite-dimensional multi-state life table. The model has transition probabilities and survival rates for each individual sample path. In other words, a mortality rate and a transition probability to the next state are assigned to each point in the continuum of ages and states. Then, the infinite-dimensional multi-state life table appears as an integral kernel that gives the cohort transition process.

---

[1] The emergence of resistant bacteria can render conventional antibiotics ineffective.

R. Oizumi, *Population Dynamics Based on Individual Stochasticity*,
Population Studies of Japan,
https://doi.org/10.1007/978-981-19-3548-0_2

2.1 Age-Structured Population Model with Individual Stochasticity

There are many ingredients characterizing individual organisms. Demographers and ecologists choose ingredients that are observable and strongly connected to life histories (e.g., size Lv and Pitchford (2007); Banks et al. (2009), weight, spatial coordinates Langlais (1988), growth stage, and income). Although age co-relates these ingredients, they are also associated with each other. Therefore, we employ the framework of the SDE that was described in the previous chapter to represent the age-specific growth rate of these ingredients. Let us define the state growth model for each trait. Suppose that $x \in A \subseteq \mathbb{R}^d$ are d-dimensional trait features characterizing each individual, where A is the domain of x. When $X_a \in A$ represents a value of x at age a, each element of $X_a = \left(X_a^j\right)_{1 \leq j \leq d}$ satisfies

$$\begin{aligned} dX_a^j &= g_j\left(a, X_a\right) da + \sum_{k=1}^{n} \sigma_{jk}\left(a, X_a\right) dB_a^k, \\ X_s^j &= x^j, \quad \sigma_{jk}\left(a, X_a\right) \geq 0 \quad j = 1, 2, \cdots, d. \end{aligned} \tag{2.1}$$

Each sample path satisfying the above equation represents the life history of each individual. The noise term governs the individual differences in heterogeneity.

As organisms cannot avoid death, these ingredients must vanish eventually. Therefore, we define age-/state-specific mortality as follows:

$$\mu\left(a, x\right) > 0, \quad \int_0^{\alpha} da\, \mu\left(a, x\right) = +\infty \quad \text{for all } x \in A, \tag{2.2}$$

where α denotes the maximum age. Therefore, the transition PDF of each state $K\left(a, y\right)$ is given by the solution of the following forward equation:

$$\begin{aligned} \frac{\partial}{\partial a} K\left(a, y\right) &= -H\left(a, y\right) K\left(a, y\right) \\ H\left(a, y\right) K\left(a, y\right) &:= \mu\left(a, y\right) K\left(a, y\right) \\ &\quad + \sum_j \frac{\partial}{\partial y^j}\left(g_j\left(a, y\right) K\left(a, y\right)\right) \\ &\quad - \frac{1}{2} \sum_{j=1}^{d} \sum_{j'=1}^{d} \frac{\partial^2}{\partial y^j \partial y^{j'}}\left(S_{jj'}\left(a, y\right) K\left(a, y\right)\right) \\ K\left(s, y\right) &= \delta^d\left(x - y\right). \end{aligned} \tag{2.3}$$

Here, the mortality corresponds to the aforementioned zero-order Kramers–Moyal coefficient. Let us consider the population involving these growth and death processes given by Eqs. (2.1) and (2.2), respectively. A state transition of the cohort in a sufficiently short time interval ε must obey the following continuous-time integral projection model:

$$P_{t+\varepsilon}(a+\varepsilon, y) = \int_{x\in A} dx\ P_t(a, x)\, K(a, x \to a+\varepsilon, y), \quad P_0(a, y) = p(a, y). \tag{2.4}$$

Recalling the initial condition in the transition kernel, it can be seen that the transition kernel $K(\cdot)$[2] is the same function as the solution of Eq. (2.3), which is called a fundamental solution. The reason for referring to it as a fundamental solution is that a general solution $\varphi(a, y)$ of Eq. (2.3) for an arbitrary initial function $\varphi(s, y) = f(s, y)$ is given by

$$\varphi(a, y) = \int_A dx\ f(s, x)\, K(s, x \to a, y). \tag{2.5}$$

Simplifying the cohort density function $P_\varepsilon(y) = P_{t+\varepsilon}(a+\varepsilon, y)$, its first derivative generates the Fokker–Planck equation identically to the derivation of the forward Kramers–Moyal equation, i.e., Eq. (1.18), such that

$$\frac{\partial}{\partial\varepsilon} P_\varepsilon(y) = -H(0, y)\, P_\varepsilon(y). \tag{2.6}$$

Note that the transition kernel $K(\cdot)$ has the zero-order Kramers–Moyal coefficient[3]:

$$\lim_{\varepsilon\downarrow 0} \int d\eta\ \frac{K(a, x \to a+\varepsilon, y+\eta)}{\varepsilon} = -\mu(a, y).$$

Owing to Eq. (2.5), the solution of Eq. (2.6) is given by

$$P_\varepsilon(y) = \int_A dx\ P_0(x)\, K(0, x \to \varepsilon, y).$$

By recalling the original parameters in the solution above, the original coordinate (t, a) is transformed by the following division into the following cases:

$$(t, a) = \begin{cases} (0, a-t) & t-a \le 0 \\ (t-a, 0) & t-a > 0 \end{cases}, \quad \varepsilon = \begin{cases} t & t-a \le 0 \\ a & t-a > 0 \end{cases}.$$

Hence, the cohort density in Eq. (2.4) is divided into two age groups as follows:

$$P_t(a, y) = \begin{cases} \int_A dx\ P_0(a-t, x)\, K(a-t, x \to a, y) & t-a \le 0 \\ \int_A dx\ P_{t-a}(0, x)\, K(0, x \to a, y) & t-a > 0 \end{cases}. \tag{2.7}$$

---

[2] This kernel plays role of an infinite-dimensional multi-state life table.

[3] We will deal with the mathematical structure of this transition kernel in the next section in this chapter.

This function also satisfies the PDE transformed by the original parameters

$$\left[\frac{\partial}{\partial t}+\frac{\partial}{\partial a}\right] P_t(a, y)=-H(a, y) P_t(a, y), \tag{2.8}$$

because

$$\frac{\partial}{\partial \varepsilon} P_{\varepsilon}(y)\Big|_{\varepsilon=0}=\left[\frac{\partial}{\partial t}+\frac{\partial}{\partial a}\right] P_t(a, y), \quad H(0, y) P_{\varepsilon}(y)\Big|_{\varepsilon=0}=H(a, y) P_t(a, y).$$

Here, $P_t(0, y)$ represents the density of newborns in state y, generated by the reproductive process. Let $F(a, y) \geq 0$ be the fertility rate at age a and state y. Then, the total number of newborns β_t satisfies the following integral equation:

$$\beta_t=\int_0^{\infty} da \int_A dy\, F(a, y) P_t(a, y). \tag{2.9}$$

In addition to Eq. (2.9), we assume that the neonatal state distribution is fixed as $\nu(y) \geq 0$, i.e.

$$P_t(0, y)=\beta_t \nu(y), \quad \int_A dy\, \nu(y)=1. \tag{2.10}$$

In general, the newborn state is influenced by the maternal state, e.g., heredity, habitat, and nutritional status. For this reason, each fertility may not be independent of the neonatal state distribution. For simplicity, we adopt the assumption given by Eq. (2.10) for the analysis in this book. From Eqs. (2.4) and (2.10), Eq. (2.7) is rewritten as

$$P_t(a, y)=\begin{cases}\int_A dx\, p(a-t, x) K(a-t, x \rightarrow a, y) & t-a \leq 0 \\ \beta_{t-a} \int_A dx\, \nu(x) K(0, x \rightarrow a, y) & t-a>0\end{cases}. \tag{2.11}$$

By substituting Eq. (2.11) into Eq. (2.9), we obtain the following renewal equation:

$$\beta_t=G_t+\int_0^t da\, \beta_{t-a} \int_A dy \int_A dx\, \nu(x) K(0, x \rightarrow a, y) F(a, y), \tag{2.12}$$

where

$$G_t:=\int_t^{\infty} da \int_A dy \int_A dx\, p(a-t, x) K(a-t, x \rightarrow a, y) F(a, y).$$

In addition to conventional age-structured models such as the McKendrick equation[4], this renewal equation implies that the total number of newborns β_t governs this

[4] In the McKendrick equation, the renewal process has the same mathematical structure as Eq. (3.7). Accordingly, the renewal equation is represented by a convolution for age (Cf. Metz and Diekmann (1986); Iannelli (1995); Inaba (2017)).

system. If the neonatal state distribution for state x depends on the maternal fertility $F(a, x, y)$, β_t becomes a function for x. The integrand of the second term, $K \times F$, on the right-hand side of Eq. (2.12) describes the expectation of the number of offspring that an individual growing from initial state x to age a and state y reproduces.

2.2 Path Integral Representation and Classical Life Schedule

Before solving the renewal equation, this section deals with the mathematical structure of the transition kernel. This function is supposed to contain the growth history of individuals and the survivorship for each individual. By analyzing this function, we can obtain a crucial representation that connects the life history of individuals with the population dynamics.

2.2.1 *Path Integral Representation and Analytical Mechanics in Demography*

First, let us select the Brownian motion as the most typical example. It plays a central role in characterizing SDEs, as mentioned previously. The Brownian motion B_t and its difference $B_t - B_s$ $(t > s)$ also obey the normal distribution. These properties arise because the Brownian motion is a continuous process. To justify its continuity mathematically, previous researchers have used the following relation concerning a one-dimensional Gaussian kernel[5]:

$$\mathcal{G}(s, x \to t, y) := \frac{1}{\sqrt{2\pi(t-s)}} \exp\left\{-\frac{(y-x)^2}{2(t-s)}\right\}$$

$$\mathcal{G}(t_1, x \to t_3, y) = \int_{-\infty}^{\infty} d\xi\, \mathcal{G}(t_1, x \to t_2, \xi)\, \mathcal{G}(t_2, \xi \to t_3, y) \quad t_1 < t_2 < t_3.$$

This relationship corresponds with the Chapman–Kolmogorov equation and its example. The convolution on the right-hand side of the above equation can be divided into any number of pieces, such that

$$\mathcal{G}(s, x \to t, y) = \int_{-\infty}^{\infty} \cdots \int_{-\infty}^{\infty} d\xi_1 \cdots d\xi_i \cdots d\xi_m$$
$$\mathcal{G}(s, x \to t_1, \xi_1) \cdots \mathcal{G}(t_i, \xi_i \to t_{i+1}, \xi_{i+1}) \cdots \mathcal{G}(t_m, \xi_m \to t, y),$$
$$s < \cdots t_i < \cdots < t_{i+1} < \cdots < t.$$

[5] This equation holds for any dimension d.

One can iterate this operation infinitely on all continuous points for all time t through several strict mathematical techniques.[6] Let us transform each time and state interval $t_{i+1} - t_i$ and $\xi_{i+1} - \xi_i$ into Δt_i and $\Delta \xi_i$, respectively. Focusing on the function in the exponential of the Gaussian kernel, it seems to be a Riemann sum for time t.

$$\begin{aligned} &\mathcal{G}(s, x \to t_1, \xi_1) \cdots \mathcal{G}(t_i, \xi_i \to t_{i+1}, \xi_{i+1}) \cdots \mathcal{G}(t_m, \xi_m \to t, y) \\ &= \frac{1}{Z} \exp\left\{ -\sum_{i=1}^{m} \frac{\Delta \xi_i^2}{2 \Delta t_i^2} \Delta t_i \right\} \delta(\xi_1 - x) \\ &t_0 = s, \xi_0 = x, t_{m+1} = t, \xi_{m+1} = y, \end{aligned}$$

where Z denotes the normalized constant

$$Z = \prod_i^m \sqrt{2\pi \Delta t_i}.$$

For a sufficiently small interval of Δt_i, physicists deem it to be an integral directly. Hence, the Gaussian kernel is rewritten as

$$\begin{aligned} \mathcal{G}(s, x \to t, y) &= \lim_{m \uparrow \infty} \lim_{\Delta t \downarrow 0} \int_{-\infty}^{\infty} \cdots \int_{-\infty}^{\infty} \prod_{i=1}^{m} \frac{d\xi_i}{Z} \exp\left\{ -\sum_i \frac{\Delta \xi_i^2}{2\Delta t_i^2} \Delta t_i \right\} \delta(\xi_1 - x) \\ &= \int_{X_s = x}^{X_t = y} \mathcal{D}(X) \frac{1}{Z} \exp\left\{ -\frac{1}{2} \int_s^t d\tau \left(\frac{dX_\tau}{d\tau} \right)^2 \right\}. \end{aligned}$$

This new form is called "path integral representation" because each integral describes a summation over all probabilities regarding pathways for the Brownian motion. However, this representation has several inconsistencies, e.g., the time derivative of X_τ does not match with its mathematical sense because the Brownian motion is not differentiable with probability one. Despite this inconsistency, it is known that various theorems regarding stochastic analysis are consistent and derivable from the interpretation of path integrals. As this book emphasizes demographic intuition rather than mathematical justification, we use the path integral representation to systematize demographic theory.

Now, for the general growth model given by Eq. (2.1), the infinitesimal difference of the column trait vector $dX_a := \left(dX_a^j\right)_{1 \le j \le d}$ follows a d-dimensional multi-variate normal distribution such that $g(a, X_a)\, da = (g(a, X_a)\, da)_{1 \le j \le d}$ and $\mathbf{S}(a, X_a)\, da := \left(S_{jj'}(a, X_a)\, da\right)_{1 \le j, j' \le d}$ give each mean growth rate and its covariance matrix, respectively (Cf. Eqs. (1.19) and (1.20)). Hence, a piece of the path integral is extended to

$$Q(a_i, \xi_i \to a_i + \Delta a_i, \xi_{i+1}) = \frac{1}{(2\pi \Delta a_i)^{\frac{d}{2}} \det\{\mathbf{S}(a_i, \xi_i)\}^{\frac{1}{2}}} \times$$

[6] It is called the Kolmogorov extension theorem Øksendal (2003)

$$\exp\left\{ -\frac{1}{2}\left(\Delta\xi_i - g\left(a_i,\xi_i\right)\Delta a_i\right)^{\top}\mathbf{S}^{-1}\left(a_i,\xi_i\right)\Delta a_i^{-1} \times \left(\Delta\xi_i - g\left(a_i,\xi_i\right)\Delta a_i\right)\right\}$$

$$= \frac{1}{\left(2\pi\Delta a_i\right)^{\frac{d}{2}}\det\left\{\mathbf{S}\left(a_i,\xi_i\right)\right\}^{\frac{1}{2}}}\times \exp\left\{-\frac{1}{2}\left(\frac{\Delta\xi_i}{\Delta a_i} - g\left(a_i,\xi_i\right)\right)^{\top}\mathbf{S}^{-1}\left(a_i,\xi_i\right) \times \left(\frac{\Delta\xi_i}{\Delta a_i} - g\left(a_i,\xi_i\right)\right)\Delta a_i\right\}.$$

Furthermore, as each state of traits has its own mortality, we should consider survivorship for a state x. Survivorship $l\left(a,x\right)$ and mortality $\mu\left(a,x\right)$ have the following relationship:

$$\frac{1}{l\left(a,x\right)}\frac{\partial}{\partial a}l\left(a,x\right) = -\mu\left(a,x\right) \quad l\left(s,x\right) = 1.$$

Therefore, the survivorship becomes

$$l\left(a,x\right) = \exp\left\{-\int_s^a d\tau\,\mu\left(\tau,x\right)\right\}.$$

In a sufficiently small age interval $\Delta a_i = a_{i+1} - a_i$, the following approximation of this function is established:

$$\exp\left\{-\int_{a_i}^{a_{i+1}} d\tau\,\mu\left(\tau,x\right)\right\} \approx \exp\left\{-\mu\left(a_i,x\right)\Delta a_i\right\}. \tag{2.13}$$

By incorporating this result with the piece of the path integral, the transition rate including the mortality consists of

$$K\left(a_i,\xi_i \to a_i+\Delta a_i,\xi_{i+1}\right) = \exp\left\{-\mu\left(a_i,\xi_i\right)\Delta a_i\right\} Q\left(a_i,\xi_i \to a_i+\Delta a_i,\xi_{i+1}\right)$$
$$= \frac{1}{\left(2\pi\Delta a_i\right)^{\frac{d}{2}}\det\left\{\mathbf{S}\left(a_i,\xi_i\right)\right\}^{\frac{1}{2}}}\times \exp\left\{-\frac{1}{2}\left(\frac{\Delta\xi_i}{\Delta a_i} - g\left(a_i,\xi_i\right)\right)^{\top}\mathbf{S}^{-1}\left(a_i,\xi_i\right) \times \left(\frac{\Delta\xi_i}{\Delta a_i} - g\left(a_i,\xi_i\right)\right)\Delta a_i - \mu\left(a_i,\xi_i\right)\Delta a_i\right\}. \tag{2.14}$$

Thus, the path integral representation corresponding with the fundamental solution given by Eq. (2.3) is written as

$$K(s, x \to a, y) = \lim_{m\uparrow\infty} \lim_{\Delta a\downarrow 0} \int_{\mathbb{R}^d} \cdots \int_{\mathbb{R}^d} \prod_{i=1}^{m} \frac{d\xi_i}{Z} K(a_i, \xi_i \to a_i + \Delta a_i, \xi_{i+1}) \delta^d(\xi_1 - x)$$
$$= \int_{X_s=x}^{X_a=y} \mathcal{D}(X) \frac{1}{Z} \exp\left\{\int_s^a d\tau \mathcal{L}\left(\tau, \dot{X}_\tau, X_\tau\right)\right\} \tag{2.15}$$
$$\mathcal{L}\left(a, \dot{X}_a, X_a\right) := -\frac{1}{2}\left(\frac{dX_a}{da} - g(a, X_a)\right)^{\top} \mathbf{S}^{-1}(a, X_a)\left(\frac{dX_a}{da} - g(a, X_a)\right) - \mu(a, X_a), \tag{2.16}$$

where Z and $\dot{X}_a$ represent the normalized constant for Q and the derivative with respect to age $\dot{X}_a = \frac{dX_a}{da}$, respectively. Note that the integration interval becomes the real d-space owing to the domain of the difference $dX_a \in \mathbb{R}^d$ at a sufficiently short time. The path integral representation shows that the connection of the SDE model generates life histories of individuals to the cohort dynamics explicitly. Some readers may want to verify whether this representation gives the Fokker–Planck equation, i.e., Eq .(2.3). As the fundamental solution obeys the multi-variate normal distribution for a sufficiently short time (recall the derivation from the Kramers–Moyal equation), we substitute Eq. (2.14) into the Chapman–Kolmogorov equation

$$K(s, x \to a + \varepsilon, y) = \int_{\mathbb{R}^d} d\eta\, K(s, x \to a, y - \eta) K(a, y - \eta \to a + \varepsilon, y)$$
$$= \int_{\mathbb{R}^d} d\eta\, K(s, x \to a, y - \eta) \exp\{-\mu(a, y - \eta)\varepsilon\}$$
$$\times Q(a, y - \eta \to a + \varepsilon, y).$$

In addition, by using the relationship

$$\exp\{-\mu(a, y - \eta)\varepsilon\} = 1 - \mu(a, y - \eta)\varepsilon, \quad \varepsilon^2 = 0,$$

we can verify whether Eq. (2.3) can be derived from the Kramers–Moyal expansion.

The integral in the exponential and its integrand are referred to as the action and Lagrangian, respectively. They play a significant role in the statistics of the growth curve. Although there are numerous growth curves from x at age s to y at age a, let us consider the most frequently realized curve. Because it must pass the point minimizing the Lagrangian, the small change in the Lagrangian becomes zero for all ages.

$$\delta \int_s^a d\tau \mathcal{L}\left(\tau, \dot{X}_\tau, X_\tau\right) = \int_{\tau=s}^{\tau=a} d\tau \left[\mathcal{L}\left(\tau, \dot{X}_\tau + \delta\dot{X}_\tau, X_\tau + \delta X_\tau\right) - \mathcal{L}\left(\tau, \dot{X}_\tau, X_\tau\right)\right] = 0$$

Here, expand the Lagrangian into the first-order Taylor series

$$\mathcal{L}\left(\tau, \dot{X}_\tau + \delta\dot{X}_\tau, X_\tau + \delta X_\tau\right) \approx \mathcal{L}\left(\tau, \dot{X}_\tau, X_\tau\right) + \sum_j \frac{\partial}{\partial \dot{X}_\tau}\mathcal{L}\left(\tau, \dot{X}_\tau, X_\tau\right)\delta\dot{X}_\tau^j + \sum_j \frac{\partial}{\partial X_\tau^j}\mathcal{L}\left(\tau, \dot{X}_\tau, X_\tau\right)\delta X_\tau^j.$$

Then, we have

$$\sum_j \int_{\tau=s}^{\tau=a} d\tau \left[\frac{\partial}{\partial \dot{X}_\tau^j}\mathcal{L}\left(\tau, \dot{X}_\tau, X_\tau\right)\delta\dot{X}_\tau^j + \frac{\partial}{\partial X_\tau^j}\mathcal{L}\left(\tau, \dot{X}_\tau, X_\tau\right)\delta X_\tau^j\right] = 0.$$

As $\delta\dot{X}_\tau^j = \frac{d}{d\tau}\delta X_\tau^j$, we perform integration by parts on the first term, and it becomes

$$\sum_j \int_{\tau=s}^{\tau=a} d\tau \left[\frac{\partial}{\partial X_\tau^j} - \frac{d}{d\tau}\frac{\partial}{\partial \dot{X}_\tau^j}\right]\mathcal{L}\left(\tau, \dot{X}_\tau, X_\tau\right)\delta X_\tau^j + \sum_j \left[\frac{\partial}{\partial \dot{X}_\tau^j}\mathcal{L}\left(\tau, \dot{X}_\tau, X_\tau\right)\delta X_\tau^j\right]_{\tau=s}^{\tau=a}.$$

As both endpoints (x, y) are fixed, δX_a and δX_s should be zero. Therefore, the growth curve satisfies

$$\left[\frac{\partial}{\partial X_\tau^j} - \frac{d}{d\tau}\frac{\partial}{\partial \dot{X}_\tau^j}\right]\mathcal{L}\left(\tau, \dot{X}_\tau, X_\tau\right) = 0 \quad X_a = y, X_s = x, \tag{2.17}$$

because δX_τ^j is an arbitrary variational. This equation is referred to as the Euler–Lagrange equation in analytical mechanics, which deals with such problems in physics. It gives a growth curve called a classical curve. Paths along this curve have a high probability in all curves having fixed endpoints. Notably, the mortality affects the observed growth curve; this is a characteristic of the individual stochasticity. The reason is that paths through a higher mortality state have a low probability of reaching the endpoint. In the deterministic life history $\mathbf{S}^{-1} \to 0$, because every path except the path

$$\frac{dX_a}{da} - g\left(a, X_a\right),$$

has zero probability, the mortality does not affect the growth curve. From another point of view, it seems to be the "principle of least action" in demography.

2.2.2 *Lagrangian and Hamiltonian Expressions in Path Integral*

If readers know the fundamentals of physics, the Hamiltonian corresponding to the Lagrangian may be of interest to them. Indeed, standard physics textbooks define the Lagrangian as follows:

$$\mathcal{L}\left(\tau, \dot{X}_\tau, X_\tau\right) = -\mathrm{i}q^{\dagger\top}\dot{X}_\tau - \mathcal{H}\left(\tau, X_\tau, -\mathrm{i}q^\dagger\right), \quad \mathrm{i}: \text{the imaginary unit}, \tag{2.18}$$

where the second term on the right-hand side is the Hamiltonian. Further, $q^{\dagger} := \left(q_j^{\dagger}\right)_{1\le j\le d}$ consists of the following solutions:

$$\dot{X}_\tau^j = \mathrm{i}\frac{\partial}{\partial q^j}\mathcal{H}\left(\tau, X_\tau, -\mathrm{i}q\right)\Big|_{q=q^{\dagger}}.$$

Equation (2.18) is called the Legendre transformation. Next, let us verify this relationship. It can justify not only the analysis by the Lagrangian, but also a similarity to several approaches in physics. Consider the Fourier transformation of the transition kernel, such that

$$\hat{K}\left(a_i, \xi_i \to a_{i+1}, q_i\right) := \frac{1}{(2\pi)^{\frac{d}{2}}}\int_{\mathbb{R}^d} d\xi_{i+1} \exp\left\{\mathrm{i}q_i^{\top}\xi_{i+1}\right\} K\left(a_i, \xi_i \to a_{i+1}, \xi_{i+1}\right).$$

Setting new variables

$$\begin{aligned}\zeta &:= \left(\zeta^j\right)_{1\le j\le d}, \quad \zeta^j := \frac{\xi_{i+1}^j - \xi_i^j}{\Delta a_i} - g_j\left(a_i, \xi_i\right), \quad \mathbf{S} = \mathbf{S}\left(a_i, \xi_i\right),\\ g_i &:= g\left(a_i, \xi_i\right), \quad \mu_i := \mu\left(a_i, \xi_i\right),\end{aligned} \tag{2.19}$$

the Fourier transformation becomes

$$\begin{aligned}\hat{K}\left(a_i, \xi_i \to a_{i+1}, q_i\right) &= \frac{1}{(2\pi)^{\frac{d}{2}}(2\pi)^{\frac{d}{2}}\det\{\mathbf{S}\}^{\frac{1}{2}}}\int_{\mathbb{R}^d}\frac{d\zeta}{(\Delta a_i)^d}\\ &\times \exp\left\{-\frac{1}{2}\zeta^{\top}\mathbf{S}^{-1}\zeta\Delta a_i + \mathrm{i}q_i^{\top}\zeta\Delta a + \mathrm{i}q_i^{\top}\left(\xi_i + g_i\Delta a_i\right) - \mu_i\Delta a_i\right\}\\ &= \frac{1}{(2\pi)^{\frac{d}{2}}(2\pi)^{\frac{d}{2}}\det\{\mathbf{S}\}^{\frac{1}{2}}}\int_{\mathbb{R}^d}\frac{d\zeta}{(\Delta a_i)^d}\\ &\times \exp\left\{-\frac{1}{2}\left(\mathbf{S}^{-\frac{1}{2}}\zeta + \mathrm{i}\mathbf{S}^{\frac{1}{2}}q_i\right)^{\top}\left(\mathbf{S}^{-\frac{1}{2}}\zeta + \mathrm{i}\mathbf{S}^{\frac{1}{2}}q_i\right)\Delta a_i - \frac{1}{2}q_i^{\top}\mathbf{S}q_i\Delta a_i\right.\\ &\left. + \mathrm{i}q_i^{\top}\left(\xi_i + g_i\Delta a_i\right) - \mu_i\Delta a_i\right\}.\end{aligned}$$

Here, we change the variables again

$$\eta = \left(\mathbf{S}^{-\frac{1}{2}}\zeta + \mathrm{i}\mathbf{S}^{\frac{1}{2}}q_i\right)\sqrt{\Delta a_i}.$$

Because the Jacobian is computed as $\det\{\mathbf{S}\}^{\frac{1}{2}}\left(\Delta a_1\right)^{\frac{d}{2}}$, we perform the multi-variate Gaussian integral

$$\begin{aligned}\hat{K}\left(a_i, \xi_i \to a_{i+1}, q_i\right) &= \frac{\exp\left\{\mathrm{i}q_i^{\top}\left(\xi_i + g\Delta a_i\right) - \frac{1}{2}q_i^{\top}\mathbf{S}q_i\Delta a_i - \mu_i\Delta a_i\right\}}{(2\pi)^{\frac{d}{2}}\left(2\pi\Delta a_i\right)^{\frac{d}{2}}}\\ &\times \int_{\mathbb{R}^d} d\eta \exp\left\{-\frac{1}{2}\eta^{\top}\eta\right\},\end{aligned}$$

and we get

$$\hat{K}\left(a_i, \xi_i \to a_{i+1}, q_i\right) = \frac{1}{\left(2\pi\,\Delta a_i\right)^{\frac{d}{2}}} \exp\left\{ \mathrm{i}q_i^{\top}\xi_i + \mathrm{i}q^{\top}g_i\Delta a_i - \frac{1}{2}q_i^{\top}\mathbf{S}q_i\Delta a_i - \mu_i\Delta a_i \right\}.$$

By the inverse transform of the above equation, we have another expression of the path integral piece

$$K\left(a_i, \xi_i \to a_{i+1}, \xi_{i+1}\right) = \frac{1}{(2\pi)^{\frac{d}{2}}} \int_{\mathbb{R}^d} dq_i \exp\left\{-\mathrm{i}q_i\xi_{i+1}\right\} \hat{K}\left(a_i, \xi_i \to a_{i+1}, q_i\right) =$$

$$\frac{1}{(2\pi)^{\frac{d}{2}}} \int_{\mathbb{R}^d} dq_i \frac{1}{\left(2\pi\,\Delta a_i\right)^{\frac{d}{2}}} \exp\left\{ \left[-\mathrm{i}q_i^{\top}\left(\frac{\xi_{i+1}-\xi_i}{\Delta a_i}\right) + \mathrm{i}q_i^{\top}g_i - \frac{1}{2}q_i^{\top}\mathbf{S}q_i - \mu_i \right]\Delta a_i \right\}.$$

This result constitutes another form of the path integral as follows:

$$K\left(s, x \to a, y\right) = \int \mathcal{D}(q) \int_{X_s=x}^{X_a=y} \mathcal{D}(X) \frac{1}{Z_1} \exp\left\{ \int_s^a d\tau \left[-\mathrm{i}q_\tau^{\top}\dot{X}_\tau - H\left(\tau, X_\tau, -\mathrm{i}q_\tau\right) \right] \right\} \tag{2.20}$$

$$Z_1 := \lim_{m\uparrow\infty}\lim_{\Delta a\downarrow 0} (2\pi)^{dm}(\Delta a)^{\frac{dm}{2}}$$

$$H\left(\tau, X_\tau, -\mathrm{i}q_\tau\right) = \mu\left(\tau, X_\tau\right) - \sum_{j=1}^{d} \mathrm{i}q_\tau^j g_j\left(\tau, X_\tau\right) + \frac{1}{2}\sum_{j=1}^{d}\sum_{j'=1}^{d} q_\tau^j q_\tau^{j'} S_{jj'}\left(\tau, X_\tau\right). \tag{2.21}$$

The function $H\left(a, X_a, -\mathrm{i}q_a\right)$ is associated with the differential operator $H\left(a, y\right)$ in Eq. (2.3). Specifically, it is shown that each frequency $\mathrm{i}q_\tau$ corresponds with the differential operator with respect to y in Eq. (2.3) such that

$$-\mathrm{i}q_\tau^j \to \frac{\partial}{\partial y^j}.$$

If this function is shown to satisfy the Legendre transformation, it becomes the Hamiltonian corresponding to the Lagrangian in the path integral. Let us verify this relationship.

Suppose that a vector $q_\tau^{\dagger} \in \mathbb{R}^d$ consists of the solution of

$$\begin{aligned} \dot{X}_\tau^j &= \mathrm{i}\frac{\partial}{\partial q^j} H\left(\tau, X_\tau, -\mathrm{i}q\right)\Big|_{q=q_\tau^{\dagger}} \\ &= g_j\left(\tau, X_\tau\right) + \frac{1}{2}\mathrm{i}\sum_{j'=1}^{d} q_\tau^{j'\dagger} S_{j'j}\left(\tau, X_\tau\right) + \frac{1}{2}\mathrm{i}\sum_{j'=1}^{d} q_\tau^{j'\dagger} S_{jj'}\left(\tau, X_\tau\right) \\ &= g_j\left(\tau, X_\tau\right) + \mathrm{i}\sum_{j'=1}^{d} q_\tau^{j'\dagger} S_{jj'}\left(\tau, X_\tau\right) \quad \Leftarrow \text{because of a property of the symmetric matrix } \mathbf{S}. \end{aligned}$$

Accordingly, we obtain

$$\mathrm{i}q_\tau^\dagger = \mathbf{S}^{-1}(\tau, X_\tau)\left(\dot{X}_\tau - g(\tau, X_\tau)\right). \tag{2.22}$$

Rewriting $H(\tau, X_\tau, \mathrm{i}q_\tau)$ in quadratic form and substituting Eq. (2.22) into Eq. (2.21), we can prove the relation to the Legendre transformation as follows:

$$\begin{aligned}
\mathcal{L}\left(\tau, \dot{X}_\tau, X_\tau\right) &= -\mathrm{i}q_\tau^{\dagger\top}\dot{X}_\tau - H\left(\tau, X_\tau, -\mathrm{i}q_\tau^\dagger\right) \\
&= -\mathrm{i}q_\tau^{\dagger\top}\dot{X}_\tau - \mu(\tau, X_\tau) + \mathrm{i}q_\tau^{\dagger\top} g(\tau, X_\tau) - \frac{1}{2}q_\tau^{\dagger\top}\mathbf{S}(\tau, X_\tau)\,q_\tau^\dagger \\
&= -\mathrm{i}q_\tau^{\dagger\top}\left(g(\tau, X_\tau) + \mathbf{S}(\tau, X_\tau)\,\mathrm{i}q_\tau^\dagger\right) - \mu(\tau, X_\tau) + \mathrm{i}q_\tau^{\dagger\top} g(\tau, X_\tau) \\
&\quad + \frac{1}{2}\mathrm{i}q_\tau^{\dagger\top}\mathbf{S}(\tau, X_\tau)\,\mathrm{i}q_\tau^\dagger \\
&= -\frac{1}{2}\mathrm{i}q_\tau^{\dagger\top}\mathbf{S}(\tau, X_\tau)\,\mathrm{i}q_\tau^\dagger - \mu(\tau, X_\tau) \\
&= -\frac{1}{2}\left(\dot{X}_\tau - g(\tau, X_\tau)\right)^\top \mathbf{S}^{-1}(\tau, X_\tau)\left(\dot{X}_\tau - g(\tau, X_\tau)\right) - \mu(\tau, X_\tau).
\end{aligned}$$

Thus,

$$H(\tau, X_\tau, -\mathrm{i}q_\tau) = \mathcal{H}(\tau, X_\tau, -\mathrm{i}q).$$

Therefore, we refer to $H(\cdot)$ as the Fokker–Planck Hamiltonian. Hence, it is shown that the analogy to analytical mechanics can be used to consider life histories in demography. The Hamiltonian in physics has the base unit of energy and is represented by the summation of kinetic energy (growth process) and potential energy (mortality), which seems to be the total energy of a cohort life history. However, the energy is not conserved because the Hamiltonian depends on age in general. Hence, organisms are in an open system. This analogy is mentioned again in the chapter 4. If we substitute Eq. (2.15) into Eq. (2.4), it is called the Lagrangian expression of the path integral, and if the other expression, i.e., Eq. (2.15) is adopted, it is called the corresponding Hamiltonian expression. Both the path integral representations are equivalent to the PDE model given by Eq. (2.8). Accordingly, we must adjust an appropriate representation for each research subject.

2.3 Characteristic Function and Eigenfunctions

We demonstrated that the integral projection model in continuous time incorporates an SDE modeling life histories using the path integral representation. We must consider the renewal process to deal with the total population dynamics. This section treats the asymptotic behavior of the *integral projection model* (IPM) of Eq. (2.4) under the renewal condition given by Eq. (2.9) from the viewpoint of stochastic analysis.

2.3.1 *Characteristic Function and Asymptotic Population Behavior*

Let us recall the renewal equation given by Eq. (3.7). It is mathematically identical to age-structured models because their state variables vanish. Hence, we can apply the conventional analytical method via the Laplace transform. Let $\hat{f}_r$ be a Laplace transform of a function $f(t)$ with respect to time t such that

$$\hat{f}_r := \int_0^\infty dt\ \exp\{-rt\} f(t).$$

Then, the Laplace transform of Eq. (3.7) becomes

$$\hat{\beta}_r = \hat{G}_r + \hat{\beta}_r \psi_r \tag{2.23}$$

$$\psi_r : = \int_0^\alpha da\ \exp\{-ra\} \int_A dy \int_A dx\ \nu(x) K(0, x \to a, y) F(a, y), \tag{2.24}$$

by using a formula of the relationship between the transform and convolution.[7] By solving this algebraic equation with respect to $\hat{\beta}_r$, we obtain

$$\hat{\beta}_r = \frac{\hat{G}_r}{1-\psi_r} = \frac{\hat{G}_r \psi_r}{1-\psi_r} + \hat{G}_r.$$

From the inverse transform, the total number of births β_t is expanded into the following residues:

$$\beta_t = \lim_{\omega \uparrow \infty} \frac{1}{2\pi \mathrm{i}} \int_{c-\mathrm{i}\omega}^{c+\mathrm{i}\omega} dr\ \exp\{rt\} \hat{\beta}_r = -\sum_{k=0}^{\infty} \left.\frac{\hat{G}_r}{\frac{d}{dr}\psi_r}\right|_{r=r_k} \exp\{r_k t\} \quad \mathrm{Re}\left[r_k\right] \geq \mathrm{Re}\left[r_{k+1}\right]. \tag{2.25}$$

Then, all the residues satisfy

$$\psi_{r_k} = 1. \tag{2.26}$$

Thus, the above equation represents the characteristic equation in this system. Eq. (2.25) is known as the Hertz–Herglotz expansion, which does not always converge Feller (1941). Alternatively, the following theorem ensures the asymptotic behavior of Eq. (2.25).

$$\beta_t = -\left.\frac{\hat{G}_r}{\frac{d}{dr}\psi_r} \exp\{rt\}\right|_{r=r_0} (1 + O(\exp\{-\delta t\})) \quad \delta > 0.$$

---

[7] $\widehat{f * g}_r = \hat{f}_r \hat{g}_r$

This theorem, called the Sharp–Lotka–Feller theorem, implies that the dominant residue governs the system.

Furthermore, the characteristic root r_0 is always real and single. Regarding the behavior of the renewal equation such as Eq. (3.7), there are various mathematical discussions; however, they are beyond the scope of this book. Interested readers may refer to several books on age-structured population models, which provide further details (e.g., Inaba (2017)). Here, we recognize this theorem as an a priori fact and show that r_0 is the dominant root. First, ψ_r monotonically decreases in r because

$$\frac{d}{dr}\psi_r = -\int_0^\infty da\, a \exp\{-ra\} \int_A dy \int_A dx\, \nu(x)\, K(0, x \to a, y)\, F(a, y).$$

Hence, the real number r_0 as the solution of Eq. (2.26) is unique. Second, let us verify that r_0 is dominant. Assume that a characteristic root r_k of Eq. (2.26) is $\mathrm{Re}[r_k] > r_0$. As r_0 and r_k satisfy

$$\begin{aligned} 1 = \psi_{r_0} = \psi_{r_k} = &\int_0^\infty da\, \exp\{-\mathrm{Re}[r_k]a\} \cos(\mathrm{Im}[r_k]a) \\ &\times \int_A dy \int_A dx\, \nu(x)\, K(0, x \to a, y)\, F(a, y), \end{aligned}$$

the integral on the right-hand side has an upper bound, such that

$$\begin{aligned} &\int_0^\infty da\, \exp\{-\mathrm{Re}[r_k]a\} \cos(\mathrm{Im}[r_k]a) \int_A dy \int_A dx\, \nu(x)\, K(0, x \to a, y)\, F(a, y) \\ &< \int_0^\infty da\, \exp\{-\mathrm{Re}[r_k]a\} \int_A dy \int_A dx\, \nu(x)\, K(0, x \to a, y)\, F(a, y). \end{aligned}$$

Because of the monotonicity for ψ_r in r, we get

$$r_0 > \mathrm{Re}[r_k].$$

This contradicts the assumption for r_k; therefore, r_0 is the dominant characteristic root. Finally, the population dynamics given by Eq. (2.11) becomes

$$\begin{aligned} P_t(a, y) &= \beta_{t-a} \int_A dx\, \nu(x)\, K(0, x \to a, y) \\ &= -\sum_{k=0}^{\infty} \left.\frac{\hat{G}_r}{\frac{d}{dr}\psi_r}\right|_{r=r_k} \exp\{r_k(t-a)\} \int_A dx\, \nu(x)\, K(0, x \to a, y) \quad t - a > 0, \end{aligned} \tag{2.27}$$

in the asymptotic sense. After a sufficiently long time, this system converges to the Malthusian growth given by the dominant characteristic root r_0 as follows:

$$P_t(a, y) \approx -\frac{\hat{G}_r}{\frac{d}{dr}\psi_r} \exp\{r(t-a)\}\Big|_{r=r_0} \int_A dx\, \nu(x) K(0, x \to a, y) \quad t \to \infty.$$

2.3.2 *Eigenfunctions and Feynman–Kac Formula*

Typical age-structured models such as the McKendrick equation have an eigenfunction expansion as their asymptotic solution. Notably, the eigenfunction and adjoint eigenfunction corresponding to the dominant characteristic root are referred to as the stable age distribution and reproductive value, respectively. The former gives an age distribution in the steady state, while the latter serves as a significant index for the evolution of life history. We intend to identify such mathematical structures in eigenfunctions in our system herein.

As shown in Eq. (2.25), the solution of Eq. (2.8) is separable.

$$P_t(a, y) = \exp\{rt\} w(a, y).$$

By substituting this into Eq. (2.8), we obtain

$$rw(a, y) = -\left[\frac{\partial}{\partial a} + H(a, y)\right] w(a, y) \quad w(0, y) = w(0)\nu(y). \tag{2.28}$$

Suppose that

$$\mathrm{w}(a, y) = \exp\{-ra\}\phi(a, y).$$

Equation (2.28) arrives at Eq. (2.3); however, the initial state is given by $w(0)\nu(y)$. Recalling the description of Eq. (2.5), the solution is

$$w(a, y) = w(0)\exp\{-ra\} \int_A dx\, \nu(x) K(0, x \to a, y). \tag{2.29}$$

Then, Eq. (2.26) determines the value of r. Next, let us derive the adjoint eigenfunction v. A functional $\langle v, w\rangle$ is defined as

$$\langle v, w\rangle := \int_0^\infty da \int_A dy\, v(a, y) w(a, y).$$

The adjoint eigenfunction represents a function that gives the following relation to the eigenfunction:

$$r\langle v, w\rangle = -\langle v, \left[\frac{\partial}{\partial a} + H(a, y)\right] w\rangle. \tag{2.30}$$

Through iterative integration by parts on the right-hand side of the above equation, we have

$$r\langle v, w\rangle = \langle\left[\frac{\partial}{\partial a} - H^*(a, y)\right]v, w\rangle - \left[\int_A dy\, v(a, y)\, w(a, y)\right]_{a=0}^{a=\alpha} \tag{2.31}$$

$$\begin{aligned} H^*(a, y)\, v(a, y) := \mu(a, y)\, v(a, y) - \sum_j g_j(a, y)\frac{\partial}{\partial y^j} v(a, y) \\ -\frac{1}{2}\sum_{j=1}^{d}\sum_{j'=1}^{d} S_{jj'}(a, y)\frac{\partial^2}{\partial y^j \partial y^{j'}} v(a, y)\,. \end{aligned} \tag{2.32}$$

Because $w(\alpha, y) = 0$ and

$$w(0, y) = \nu(y)\langle F, w\rangle,$$

the second term on the right-hand side of Eq. (2.31) is replaced by

$$\int_A dy\, v(0, y)\, w(0, y) = \int_A dy\, v(0, y)\, \nu(y)\langle F, w\rangle.$$

Thus, Eq. (2.31) is rewritten as follows:

$$r\langle v, w\rangle = \langle\left[\frac{\partial}{\partial a} - H^*(a, y)\right]v + \int_A d\xi\, v(0, \xi)\, \nu(\xi)\, F, w\rangle.$$

Consequently, the adjoint eigenfunction corresponding to arbitrary $w(a, y)$ must satisfy

$$0 = \left[\frac{\partial}{\partial a} - H^*(a, y) - r\right]v(a, y) + \int_A d\xi\, v(0, \xi)\, \nu(\xi)\, F(a, y) \quad v(\alpha, y) = 0. \tag{2.33}$$

Hence, the adjoint eigenfunction follows

$$v(a, y) = \int_A d\xi\, v(0, \xi)\, \nu(\xi)\int_a^\alpha d\tau\, \exp\{-r(\tau - a)\}\int_A d\xi\, K(a, y \to \tau, \xi)\, F(\tau, \xi)\,. \tag{2.34}$$

Instead of the derivation for the representation above, we examine the derivation of another form from the perspective of stochastic analysis. Given a function consisting of the solution of Eq. (2.1),

$$\begin{aligned} Y_\tau = v(\tau, X_\tau)\exp\left\{-\int_a^\tau d\tau_1\, \left(\mu(\tau_1, X_{\tau_1}) + r\right)\right\} \\ + \int_A d\xi\, v(0, \xi)\, \nu(\xi)\int_a^\tau d\tau_1\, F(\tau_1, X_{\tau_1})\exp\left\{-\int_a^{\tau_1} d\tau_2\, \left(\mu(\tau_2, X_{\tau_2}) + r\right)\right\}, \end{aligned}$$

its stochastic differential is

$$dY_\tau = \underbrace{d\left(v(\tau, X_\tau)\exp\left\{-\int_a^\tau d\tau_1\ \left(\mu\left(\tau_1, X_{\tau_1}\right)+r\right)\right\}\right)}_{\mathrm{I}}$$
$$+\underbrace{d\left(\int_A d\xi\ v(0,\xi)\,\nu(\xi)\int_a^\tau d\tau_1\ F\left(\tau_1, X_{\tau_1}\right)\exp\left\{-\int_a^{\tau_1} d\tau_2\ \left(\mu\left(\tau_2, X_{\tau_2}\right)+r\right)\right\}\right)}_{\mathrm{II}}.$$

From Ito's formula, i.e., Eq. (1.12), the first term I on the right-hand side generates

$$\begin{aligned}
\mathrm{I} &= dv(\tau, X_\tau)\exp\left\{-\int_a^\tau d\tau_1\ \left(\mu\left(\tau_1, X_{\tau_1}\right)+r\right)\right\} \\
&\quad + v(\tau, X_\tau)\,d\exp\left\{-\int_a^\tau d\tau_1\ \left(\mu\left(\tau_1, X_{\tau_1}\right)+r\right)\right\} \\
&= \exp\left\{-\int_a^\tau d\tau_1\ \left(\mu\left(\tau_1, X_{\tau_1}\right)+r\right)\right\}\left[\frac{\partial}{\partial\tau} - H^*(\tau, X_\tau) - r\right] v(\tau, X_\tau)\,d\tau \\
&\quad + \exp\left\{-\int_a^\tau d\tau_1\ \left(\mu\left(\tau_1, X_{\tau_1}\right)+r\right)\right\}\sum_{j=1}^{d}\sum_{k=1}^{n}\sigma_{jk}(\tau, X_\tau)\frac{\partial}{\partial X^j}v(\tau, X_\tau)\,dB_\tau^k.
\end{aligned}$$

Similarly, the second term II gives

$$\mathrm{II} = \int_A d\xi\ v(0,\xi)\,\nu(\xi)\,F(\tau, X_\tau)\exp\left\{-\int_a^\tau d\tau_1\ \left(\mu\left(\tau_1, X_{\tau_1}\right)+r\right)\right\}d\tau.$$

Putting the pieces together, we get

$$\begin{aligned}
dY_\tau &= \mathrm{I} + \mathrm{II} \\
&= \exp\left\{-\int_a^\tau d\tau_1\ \left(\mu\left(\tau_1, X_{\tau_1}\right)+r\right)\right\}\left(\left[\frac{\partial}{\partial\tau} - H^*(\tau, X_\tau) - r\right] v(\tau, X_\tau)\right. \\
&\quad \left. + \int_A d\xi\ v(0,\xi)\,\nu(\xi)\,F(\tau, X_\tau)\right) d\tau \\
&\quad + \exp\left\{-\int_a^\tau d\tau_1\ \left(\mu\left(\tau_1, X_{\tau_1}\right)+r\right)\right\}\sum_{j=1}^{d}\sum_{k=1}^{n}\sigma_{jk}(\tau, X_\tau)\frac{\partial}{\partial X^j}v(\tau, X_\tau)\,dB_\tau^k.
\end{aligned}$$

Now, $v(\tau, X_\tau)$ is a solution of Eq. (2.33), and the first term on the right-hand side of the above equation is zero. Moreover, owing to the property of Brownian motion given by Eq. (1.3), the expectation of the second term is zero.

$$\mathbb{E}\left[\exp\left\{-\int_a^\tau d\tau_1\ \left(\mu\left(\tau_1, X_{\tau_1}\right)+r\right)\right\}\sum_{j=1}^{d}\sum_{k=1}^{n}\sigma_{jk}(\tau, X_\tau)\frac{\partial}{\partial X^j}v(\tau, X_\tau)\,dB_\tau^k\right] = 0.$$

Therefore

$$\mathbb{E}_{X_a=y}[Y_\tau] = \mathbb{E}_{X_a=y}\Bigg[v(\tau, X_\tau)\exp\left\{-\int_a^\tau d\tau_1\ \left(\mu\left(\tau_1, X_{\tau_1}\right)+r\right)\right\}$$
$$+\int_A d\xi\ v(0,\xi)\,\nu(\xi)\int_a^\tau d\tau_1\ F\left(\tau_1, X_{\tau_1}\right)$$
$$\times\exp\left\{-\int_a^{\tau_1} d\tau_2\ \left(\mu\left(\tau_2, X_{\tau_2}\right)+r\right)\right\}\Bigg],$$

is a candidate of the solution for Eq. (2.33). As the terminal condition gives $v(\alpha, y) = 0$, the solution is as follows:

$$v(a, y) = \int_A d\xi\ v(0,\xi)\,\nu(\xi)\,\mathbb{E}_{X_a=y}\Bigg[\int_a^\alpha d\tau\ F(\tau, X_\tau)$$
$$\times\exp\left\{-\int_a^\tau d\tau_1\ \left(\mu\left(\tau_1, X_{\tau_1}\right)+r\right)\right\}\Bigg]. \tag{2.35}$$

Compared with Eq. (2.34), it seems that the transition kernel given by Eq. (2.15) yields the statistical representation above. This representation is mathematically correct, and the kernel contains all the survivorship for each path in the mathematical structure. It is called the Feynman–Kac formula, which expresses the solution of the parabolic PDE given by Eq. (2.33) as the statistics of a stochastic process. The name of this formula comes from Feynman's work for devising the path integral representation in quantum mechanics Feynman et al. (1948) as well as Kac's work on the heat equation with the potential function Kac (1949). The interpretation of the Feynman–Kac formula gives the characteristic equation, i.e., Eq. (2.26), in another form

$$\psi_{r_k} = \int_A dy\,\mathbb{E}_{X_0=y}\left[\int_0^\alpha da\ F(a, X_a)\exp\left\{-\int_0^a d\tau\ \left(\mu\left(\tau, X_\tau\right)+r_k\right)\right\}\right]\nu(y) = 1. \tag{2.36}$$

The structure in the expectation above is identical to the characteristic equation in the McKendrick equation that the Laplace transform for the product of fertility and survivorship constitutes. The same is true for the adjoint eigenfunction; it corresponds with the reproductive value in age-structured models. Finally, we demonstrate the orthogonality between eigenfunctions and their adjoint such that

$$\langle v_k, w_{k'}\rangle = \begin{cases} 0 & k \neq k' \\ -C_k \frac{d}{dr}\psi_r\Big|_{r=r_k} & k = k' \end{cases},\quad C_k := \int_A d\xi\ v_k(0,\xi)\,\nu(\xi)\,w_k(0). \tag{2.37}$$

Here, $v_k(a, y)$ and $w_k(a, y)$ represent the adjoint and the eigenfunctions, respectively, corresponding to the k-th characteristic root r_k. Considering a functional of eigenfunctions corresponding to different characteristic roots, which follows

$$
\begin{aligned}
\langle v_k, w_{k'} \rangle &= \int_0^\infty da \int_A dy\, v_k(a, y)\, w_{k'}(a, y) \\
&= \int_0^\infty da \int_A dy \int_A d\xi\, v_k(0, \xi)\, \nu(\xi)\, w_{k'}(0) \int_a^\infty d\tau\, \exp\{-r_k(\tau - a)\} \\
&\quad \times \int_A d\xi\, K(a, y \to \tau, \xi)\, F(\tau, \xi) \exp\{-r_{k'} a\} \int_A dx\, \nu(x)\, K(0, x \to a, y) \\
&= \int_0^\infty da \int_A d\xi\, v_k(0, \xi)\, \nu(\xi)\, w_{k'}(0) \int_a^\infty d\tau\, \exp\{-r_k \tau - (r_{k'} - r_k) a\} \\
&\quad \times \int_A dx\, \nu(x)\, K(0, x \to \tau, \xi)\, F(\tau, \xi) \\
&= \int_0^\infty d\tau \int_A d\xi\, v_k(0, \xi)\, \nu(\xi)\, w_{k'}(0) \int_0^\tau da\, \exp\{-r_k \tau - (r_{k'} - r_k) a\} \\
&\quad \times \int_A dx\, \nu(x)\, K(0, x \to \tau, \xi)\, F(\tau, \xi) \\
&= \frac{-1}{r_{k'} - r_k} \int_0^\infty d\tau \int_A d\xi\, v_k(0, \xi)\, \nu(\xi)\, w_{k'}(0) \left[\exp\{-r_k \tau - (r_{k'} - r_k) a\}\right]_{a=0}^{a=\tau} \\
&\quad \times \int_A dx\, \nu(x)\, K(0, x \to \tau, \xi)\, F(\tau, \xi) \\
&= \frac{1}{r_{k'} - r_k} \int_0^\infty d\tau \int_A d\xi\, v_k(0, \xi)\, \nu(\xi)\, w_{k'}(0) \left[\exp\{-r_k \tau\} - \exp\{-r_{k'} \tau\}\right] \\
&\quad \times \int_A dx\, \nu(x)\, K(0, x \to \tau, \xi)\, F(\tau, \xi) \\
&= 0
\end{aligned}
$$

Because both r_k and $r_{k'}$ are characteristic roots of Eq. (2.26), it becomes zero. Conversely, when $k = k'$, we can verify Eq. (2.37) as well. Furthermore, $\hat{G}_r$ can be divided into a functional of the initial population and the adjoint eigenvector:

$$
\begin{aligned}
\hat{G}_r &= \int_0^\infty dt\, \exp\{-rt\} \int_t^\infty da \int_A dy \int_A dx\, p(a - t, x)\, K(a - t, x \to a, y)\, F(a, y) \\
& a - t \to \tau \\
&= \int_0^\infty dt\, \exp\{-rt\} \int_0^\infty d\tau \int_A dy \int_A dx\, p(\tau, x)\, K(\tau, x \to \tau + t, y)\, F(\tau + t, y) \\
& \tau + t \to \theta \\
&= \int_0^\infty d\tau \int_\tau^\infty d\theta \exp\{-r(\theta - \tau)\} \int_A dy \int_A dx\, p(\tau, x)\, K(\tau, x \to \theta, y)\, F(\theta, y) \\
&= \int_0^\infty d\tau \int_A dx \int_\tau^\infty d\theta\, \exp\{-r(\theta - \tau)\} \int_A dy\, K(\tau, x \to \theta, y)\, F(\theta, y)\, p(\tau, x) \\
&= \frac{\langle v, p \rangle}{\int_A d\xi\, v(0, \xi)\, \nu(\xi)\, w(0)}.
\end{aligned}
$$

Accordingly, we can rewrite the population dynamics given by Eq. (2.27) as the following eigenfunction expansion.[8]

$$P_t(a, y) = \sum_{k=0}^{\infty} \frac{\langle v_k, p \rangle}{\langle v_k, w_k \rangle} \exp\{r_k t\} w_k(a, y) \quad t - a > 0. \tag{2.38}$$

Indeed, expanding the initial population into these eigenfunctions

$$p(a, y) = \sum_{k=0}^{\infty} c_k w_k(a, y), \tag{2.39}$$

the orthogonality between the eigenfunctions given by Eq. (2.37) determines each expansion coefficient, such that

$$\langle v_k, p \rangle = c_k \langle v_k, w_k \rangle \Rightarrow c_k = \frac{\langle v_k, p \rangle}{\langle v_k, w_k \rangle}.$$

However, as mentioned previously, the series given by Eq. (2.38) does not converge and is associated with the left-hand side of Eq. (2.39) at $t = 0$ in general. Recall the Sharp–Lotka–Feller theorem. The asymptotic behavior is rewritten as

$$P_t(a, y) = \frac{\langle v_0, p \rangle}{\langle v_0, w_0 \rangle} \exp\{r_0 t\} w_0(a, y) (1 + O(\exp\{-\delta t\})) \quad \delta > 0. \tag{2.40}$$

Thus, the age–multi-state population model based on the SDE is parallel to the mathematical structure of the conventional age-structured population model. The eigenfunction corresponding to r_0, $w_0(a, y)$, gives the stable age and multi-state distribution, and its adjoint eigenfunction $v_0(a, y)$ is associated with the reproductive value, which was defined by Fisher Fisher (1930). Notably, the latter plays a significant role in the discussion on life history evolution.

2.4 Demographic Indicators

The mathematical structure of the structured population model generated by SDE life history models is identical to that of age-structured models. In the previous sections in this chapter, we showed that the statistics of stochastic life histories represented several components that determine the mathematical structure, e.g., the characteristic function and the eigenfunction. In this section, we focus on conventional demographic indices and rewrite them using the statistics of individual stochasticity.

Under individual stochasticity, organisms have different survivorship and reproductive processes depending on each path of their life history. Therefore, several

[8] If and only if all characteristic roots are not duplicated.

demographic indicators could be represented as statistics of sample-paths. Let us consider an integral of the transition kernel for y. From the Feynman–Kac formula, it implies an expectation of survivorship from age s and state x to age a, such that

$$\int_A dy\ K(s, x \to a, y) = \mathbb{E}_{X_s = y}\left[\exp\left\{-\int_s^a d\tau \mu(\tau, X_\tau)\right\}\right].$$

As for the survivorship at age a in the whole population, we can obtain it using the initial state distribution

$$\int_A dx \int_A dy\ \nu(x)\, K(0, x \to a, y) = \mathbb{E}\left[\exp\left\{-\int_0^a d\tau \mu(\tau, X_\tau)\right\}\right].$$

The life expectancy at birth is also represented by the transition kernel as follows:

$$\begin{aligned}
e_0 :&= \int_0^\infty da\ \mathbb{E}\left[a\mu(a, X_a)\exp\left\{-\int_0^a d\tau\mu(\tau, X_\tau)\right\}\right] \\
&= -\mathbb{E}\left[\int_0^\infty da\ a\frac{d}{da}\exp\left\{-\int_0^a d\tau\mu(\tau, X_\tau)\right\}\right] \\
&= -\left[\mathbb{E}\left[a\exp\left\{-\int_0^a d\tau\mu(\tau, X_\tau)\right\}\right]\right]_{a=0}^{a=\alpha} \\
&\quad + \mathbb{E}\left[\int_0^\infty da\ \exp\left\{-\int_0^a d\tau\mu(\tau, X_\tau)\right\}\right] \\
&= \mathbb{E}\left[\int_0^\alpha da\ \exp\left\{-\int_0^a d\tau\mu(\tau, X_\tau)\right\}\right] \\
&= \int_0^\alpha da \int_A dx \int_A dy\ \nu(x)\, K(0, x \to a, y).
\end{aligned}$$

The life expectancy itself is given by

$$\begin{aligned}
e_a :&= \frac{\int_A dy \int_A dx\ \nu(x)\, K(0, x \to a, y) \int_a^\alpha d\tau \int_A dz K(a, y \to \tau, z)}{\int_A dy \int_A dx\ \nu(x)\, K(0, x \to a, y)} \\
&= \frac{\int_a^\alpha d\tau \int_A dx\ \nu(x) \int_A dz\ K(0, x \to \tau, z)}{\int_A dx\ \nu(x) \int_A dy\ K(0, x \to a, y)} \\
&= \frac{\mathbb{E}\left[\int_a^\alpha d\tau\ \exp\left\{-\int_0^\tau d\tau_1\ \mu(\tau_1, X_{\tau_1})\right\}\right]}{\mathbb{E}\left[\exp\left\{-\int_0^a d\tau\ \mu(\tau, X_\tau)\right\}\right]}.
\end{aligned}$$

Regarding reproduction, the *basic reproduction number*[9] (BRN) is one of the most significant indices that determine the increase and decrease in the population. The

[9] It has the same meaning as the *net reproduction rate* in human demography. In this book, the term "basic reproduction number" is used for consistency with biological concepts.

indicator can be derived from the function constituting the Euler–Lotka equation, i.e., Eq. (2.26), directly, and it is given by

$$R_0 = \psi_r\Big|_{r=0} = \int_A dx\, \nu(x)\, \mathbb{E}_{X_0=x}\left[\int_0^\alpha da\, F(a, X_a) \exp\left\{-\int_0^a d\tau\, \mu(\tau, X_\tau)\right\}\right]$$
$$= \mathbb{E}\left[\int_0^\alpha da\, F(a, X_a) \exp\left\{-\int_0^a d\tau\, \mu(\tau, X_\tau)\right\}\right]. \tag{2.41}$$

The integrand constituting BRN can form an age structure for the reproductive age because the expectation of the integrand indicates the net reproduction distribution. Significantly, the reproductive age distribution $\mathcal{R}(a)$ follows

$$\mathcal{R}(a) := \frac{1}{R_0}\mathbb{E}\left[F(a, X_a) \exp\left\{-\int_0^a d\tau\, \mu(\tau, X_\tau)\right\}\right]. \tag{2.42}$$

The cumulant-generating function Θ_r concerning the reproductive age $a^\diamond$ consists of the Laplace transform of this distribution for a:

$$\Theta_r := \ln\{\psi_r\} = \sum_{m=1}^{\infty} \frac{(-r)^m}{m!} \langle a^\diamond \rangle^{(m)}, \tag{2.43}$$

where each cumulant is given by

$$\langle a^\diamond \rangle^{(m)} := (-1)^m \frac{d^m}{dr^m}\Theta_r\Big|_{r=0}.$$

The first- and second-order cumulants denote the expectation and variance of the age, respectively

$$\langle a^\diamond \rangle^{(1)} = \int_0^\alpha da\, a\mathcal{R}(a), \tag{2.44}$$
$$\langle a^\diamond \rangle^{(2)} = \int_0^\alpha da\, a^2\mathcal{R}(a) - \left(\int_0^\alpha da\, a\mathcal{R}(a)\right)^2. \tag{2.45}$$

Thus, several demographic indices reappear as the statistics of individual life histories.

2.5 Application to a Semelparity Model

The previous sections in this chapter focused on the systematization of linear structured population models based on individual stochasticity. Here, we clarify the effect of individual stochasticity on the population dynamics via a solvable and straightforward example. We believe that semelparous species are suitable examples for understanding the effect because they must decide their reproductive timing under uncertain life events.

For simplicity, consider a species that grows in size following the geometric Brownian motion as follows:

$$dX_a = g_0 X_a da + \sigma_0 X_a dB_a, \quad X_0 = x > 0, \tag{2.46}$$

where $g_0, \sigma_0 \geq 0$ are constant. Suppose that mortality occurs by accidental death only at a fixed rate of $\mu_0 > 0$. Meanwhile, each individual grows to the proper size $x^\diamond$ and then produces offspring while ensuring death. If $a^\diamond$ represents the age when the proper size is first reached, which is referred to as the mature age[10],

$$a^\diamond := \inf_a \left\{ a \in (0, \infty) \middle| X_a = x^\diamond \right\},$$

the mortality becomes

$$\mu_s(x) := \begin{cases} \mu_0 & a \leq a^\diamond \\ \infty & a > a^\diamond \end{cases}. \tag{2.47}$$

As reproductive behavior only occurs at the proper size $x^\diamond$ upon maturity, the fertility rate gives a boundary condition

$$F_s(x) := \begin{cases} 0 & x < x^\diamond \\ \phi(x) > 0 & x = x^\diamond \end{cases}, \tag{2.48}$$

where $\phi(x)$ denotes the size-specific litter size. By combining Eqs. (2.47) and (2.48), we model a semelparous life history. Here, note that the mature age becomes a random variable. All individuals are assumed to have an identical initial size x such that

$$\nu(y) = \delta(x - y).$$

According to the characteristic equation represented by the Feynman–Kac formula, these configurations must be expressed as follows:

$$\psi_r(x) = \begin{cases} \mathbb{E}_{X_0=x}\left[\int_0^\infty da \, \exp\{-ra\} F_s(X_a) \exp\left\{-\int_0^a d\tau \, \mu_s(X_\tau)\right\}\right] & x < x^\diamond \\ \phi(x^\diamond) & x = x^\diamond \end{cases}. \tag{2.49}$$

If $x = x^\diamond$, it is biologically nonsense; however, the individual reproduces at a rate ϕ with probability one and dies at an age greater than zero. If the life history models are autonomous systems[11] such as Eqs. (2.46) and (2.47), they have the "strong

[10] This age has the same meaning as the reproductive age in this type of semelparity.

[11] An autonomous system implies that the ingredients for the system depend on X_a only such that $g_j(a, X_a) = g_j(X_a)$, $\sigma_{jk}(a, X_a) = \sigma_{jk}(X_a)$, $\mu(a, X_a) = \mu(X_a)$, and $F(a, X_a) = F(X_a)$ in Eqs. (2.1), (2.2), and (2.9), respectively.

Markov" property. First, the Markov property describes the translational symmetry for age (time):

$$K(s, x \to s+\tau, y) = K(a, x \to a+\tau, y), \quad s \le a, \quad 0 < \tau.$$

The strong Markov property holds for the relation above even in τ as a random time. Expressly, it permits computation regarding the mature age as follows:

$$\begin{aligned}
\psi_r(x) &= \mathbb{E}_{X_0=x}\left[\int_0^\infty da\ \exp\{-ra\} F_s(X_a) \exp\left\{-\int_0^a d\tau\ \mu_s(X_\tau)\right\}\right] \\
&= \mathbb{E}_{X_0=x}\left[\int_{a^\diamond}^\infty da\ \exp\{-ra\} F_s(X_a) \exp\left\{-\int_0^a d\tau\ \mu_s(X_\tau)\right\}\right] \\
&\Leftarrow \text{becase of Eq. (73)} \\
&= \mathbb{E}_{X_0=x}\left[\exp\{-ra^\diamond\} \int_0^\infty da\ \exp\{-ra\} F_s(X_{a+a^\diamond})\right. \\
&\quad \left.\times \exp\left\{-\int_0^{a+a^\diamond} d\tau\ \mu_s(X_\tau)\right\}\right] \\
&= \mathbb{E}_{X_0=x}\left[\exp\{-(r+\mu_0)a^\diamond\}\right] \mathbb{E}_{X_0=x^\diamond}\left[\int_0^\infty da\ \exp\{-ra\} F_s(X_a)\right. \\
&\quad \left.\times \exp\left\{-\int_0^a d\tau\ \mu_s(X_\tau)\right\}\right] \\
&= \mathbb{E}_{X_0=x}\left[\exp\{-(r+\mu_0)a^\diamond\}\right] \phi(x^\diamond). \Uparrow \text{the strong Markov property.}
\end{aligned} \tag{2.50}$$

For readers who are not familiar with probability theory, such a calculation may be difficult to understand. We explain it using another approach. The transition kernel,

$$K(s, x \to a, y),$$

represents the probability density function of survivors who transition from state x at age s to state y at age a. This distribution contains the probabilities of the first paths that reach y between these ages as well as those of many recurrent paths that reach it. If the stochastic process can reach the size, its probability should consist of the first paths and recurrent paths. Furthermore, the probabilities for these paths are independent of each other in the Markovian process. For these reasons, the kernel is divided into probabilities of the first reaching paths and otherwise. Therefore,

$$K(s, x \to a, y) = \int_s^a d\tau\ Q(s, x \to \tau, x^\diamond) K(\tau, x^\diamond \to a, y),$$

where Q denotes the first-reaching-age distribution from x at age s to $x^\diamond$ for τ. Now, we require the statistics of the transition kernel consisting of the first reaching paths. Rewrite the function constituting the characteristic function as an integral representation

$$\psi_r(x) = \int_0^\infty da \int_0^\infty dy \, \exp\{-ra\} F_s(y) K_s(0, x \to a, y)$$
$$= \int_0^\infty da \, \exp\{-ra\} K\left(0, x \to a, x^\diamond\right) \phi\left(x^\diamond\right)$$

Then, we substitute the decomposition into the above equation:

$$\psi_r(x) = \int_0^\infty da \, \exp\{-ra\} \int_0^a d\tau \, \mathcal{Q}_s\left(0, x \to \tau, x^\diamond\right) K\left(\tau, x^\diamond \to a, x^\diamond\right) \phi\left(x^\diamond\right)$$
$$= \int_0^\infty d\tau \, \exp\{-r\tau\} \mathcal{Q}_s\left(0, x \to \tau, x^\diamond\right) \int_0^\infty d\tau' \, \exp\left\{-r\tau'\right\} K$$
$$\times \left(\tau, x^\diamond \to \tau + \tau', x^\diamond\right) \phi\left(x^\diamond\right)$$
$$= \int_0^\infty d\tau \, \exp\{-r\tau\} \mathcal{Q}_s\left(0, x \to \tau, x^\diamond\right) \int_0^\infty d\tau' \, \exp\left\{-r\tau'\right\}$$
$$\times \underbrace{K\left(0, x^\diamond \to \tau', x^\diamond\right)}_{\text{the strong Markov property}} \phi\left(x^\diamond\right)$$
$$= \int_0^\infty d\tau \, \exp\{-r\tau\} \mathcal{Q}_s\left(0, x \to \tau, x^\diamond\right) \int_0^\infty d\tau' \, \exp\left\{-r\tau'\right\} \mathcal{Q}_s$$
$$\times \left(0, x^\diamond \to \tau', x^\diamond\right) \phi\left(x^\diamond\right)$$
$$= \int_0^\infty d\tau \, \exp\{-r\tau\} \mathcal{Q}_s\left(0, x \to \tau, x^\diamond\right) \int_0^\infty d\tau' \, \exp\left\{-r\tau'\right\} \delta\left(\tau'\right) \phi\left(x^\diamond\right)$$
$$= \int_0^\infty d\tau \, \exp\{-r\tau\} \mathcal{Q}_s\left(0, x \to \tau, x^\diamond\right) \phi\left(x^\diamond\right).$$

As $\mathcal{Q}_s$ represents the mature age distribution in the semelparous system, it becomes a delta function that concentrates the mass at age zero,

$$\lim_{x \uparrow x^\diamond} \mathcal{Q}_s\left(0, x \to \tau', x^\diamond\right) = \delta\left(\tau'\right),$$

if we choose the initial state as the mature size $x^\diamond$.[12] Hence, the Laplace transform of the PDF for a mature age is interpreted as Eq. (2.50). According to this interpretation, the BRN in semelparity is equivalent to the product of the reaching probability and its litter size.

$$\psi_r(x)\Big|_{r=0} = \mathbb{E}_{X_0=x}\left[\exp\left\{-\mu_0 a^\diamond\right\}\right] \phi\left(x^\diamond\right) \tag{2.51}$$

Because the characteristic function leads to several remarkable demographic indices, its derivation is vital to various applications. First, let us consider an adjoint eigenfunction (Cf. Eq. (2.35)) in a general autonomous system:

[12] Owing to the definition of mortality, recurrent paths do not exist such that $K\left(0, x^\diamond \to \tau', x^\diamond\right) = \lim_{x \uparrow x^\diamond} \mathcal{Q}_s\left(0, x \to \tau', x^\diamond\right)$

$$v(a, y) = \mathbb{E}_{X_a=y}\left[\int_a^\alpha ds\ \exp\{-ra\} F(X_s) \exp\left\{-\int_0^s d\tau\ \mu(X_\tau)\right\}\right].$$

For simplicity, we set an arbitrary constant to one. Then, from the Markov property, the lifespan can move parallel to the following interval:

$$v(a, y) = \mathbb{E}_{X_0=y}\left[\int_0^{\alpha-a} ds\ \exp\{-rs\} F(X_s) \exp\left\{-\int_0^s d\tau\ \mu(X_\tau)\right\}\right].$$

Because the mature age lies in the range of zero to infinity, the adjoint eigenfunction becomes

$$\lim_{\alpha\uparrow\infty} v(a, y) = \mathbb{E}_{X_0=y}\left[\int_0^\infty ds\ \exp\{-rs\} F(X_s) \exp\left\{-\int_0^s d\tau\ \mu(X_\tau)\right\}\right] = v(y).$$

Thus, the autonomous system with the infinite maximum age has a reproductive value independent of age. The adjoint equation (see Eq. (2.33)) that the eigenfunction follows also becomes

$$0 = -\left[H^*(y) + r\right] v(y) + F(y). \tag{2.52}$$

From the same method to derive the adjoint equation, the adjoint Hamiltonian in the semelparous model is given by

$$H_s^*(y) = \mu_0 - g_0 y \frac{d}{dy} - \frac{\sigma_0^2}{2} y^2 \frac{d^2}{dy^2}. \tag{2.53}$$

It derives the equation that the adjoint eigenfunction satisfies from the equations above.

$$0 = -\left[H_s^*(y) + r\right] v(y) + F_s(y). \tag{2.54}$$

This is a boundary condition problem in differential equations, which is called the Dirichlet problem.[13] As shown in the semelparous fertility function given by Eqs. (2.48) and (2.54) implies that $v(y)$ obeys

$$\text{Eq. (79)} = \begin{cases} -\left[H_s^*(y) + r\right] v(y) = 0 & y < x^\diamond \\ v(y) = \phi(y) & y = x^\diamond \end{cases}.$$

By recalling that all individuals have an identical initial size x, the characteristic function is also the solution of the adjoint equation[14]

[13] This problem is to solve an ODE or a PDE under a specified value (or function) on their boundary.

[14] Rigorously, it is given by,

$$\psi_r(x) = \int_0^\infty dy\ v(y)\, \delta(x-y).$$

$$v(x) = \psi_r(x).$$

Let us solve Eq. (2.54) explicitly. Assume that the solution is a power of y with an arbitrary constant such as

$$v(y) = Cy^{\rho}, \quad C \neq 0.$$

Substituting this into Eq. (2.54), we get

$$\begin{aligned} 0 &= -\left[\mu_0 - g_0\rho - \frac{\sigma_0^2}{2}\rho(\rho-1) + r\right]Cy^{\rho} \\ &= \mu_0 - g_0\rho - \frac{\sigma_0^2}{2}\rho(\rho-1) + r. \end{aligned} \tag{2.55}$$

The exponent that satisfies the above equation is

$$\rho_{\mp} = \frac{1}{2}\left[\left(1 - \frac{2g_0}{\sigma_0^2}\right) \mp \sqrt{\left(1 - \frac{2g_0}{\sigma_0^2}\right)^2 + \frac{8}{\sigma_0^2}(\mu_0 + r)}\,\right].$$

Because these two exponents are capable of constituting the solution, the summation of these solutions is also the solution as follows:

$$v(y) = C_- y^{\rho_-} + C_+ y^{\rho_+}.$$

From Eq. (2.46),$v(y)$ must satisfy $\lim_{y\downarrow 0} v(y) = 0$, and the boundary condition imposes $\lim_{y\uparrow x^{\diamond}} v(y) = \phi(x^{\diamond})$. Hence, each constant should be $C_- = 0$ and $C_+ = \phi(x^{\diamond})(x^{\diamond})^{-\rho_+}$. Then, we obtain the adjoint eigenfunction in this semelparous model

$$v(y) = \phi(x^{\diamond})\left(\frac{y}{x^{\diamond}}\right)^{\rho_+}. \tag{2.56}$$

Accordingly, the characteristic function is also given by

$$\psi_r(x) = \phi(x^{\diamond})\left(\frac{x}{x^{\diamond}}\right)^{\rho_+}. \tag{2.57}$$

Now, we can solve the Euler–Lotka equation:

$$\begin{aligned} 1 &= \phi(x^{\diamond})\left(\frac{x}{x^{\diamond}}\right)^{\rho_+}\Bigg|_{r=r_0} \\ &= \exp\left\{\rho_+(r_0)\ln\left\{\frac{x}{x^{\diamond}}\right\} + \ln\{\phi(x^{\diamond})\}\right\}. \end{aligned}$$

It shows that r_0 provides the following relationship:

$$\rho_+(r_0) = \frac{\ln\{\phi(x^\diamond)\}}{\ln\left\{\frac{x^\diamond}{x}\right\}}.$$

Substituting $\rho_+(r_0)$ into Eq. (2.55), we obtain the dominant characteristic root:

$$r_0 = \rho_0\left[g_0 - \frac{\sigma_0^2}{2}(1-\rho_0)\right] - \mu_0, \quad \rho_0 := \frac{\ln\{\phi(x^\diamond)\}}{\ln\left\{\frac{x^\diamond}{x}\right\}}. \tag{2.58}$$

This is *the intrinsic rate of natural increase* (IRNI) in the semelparous modelOizumi and Takada (2013). It shows that the effect of individual stochasticity σ_0 on the population growth is strongly associated with the parameter ρ_0. In particular, the value ρ_0, which is more or less than one, determines whether σ_0 has positive or negative effects.

$$\frac{\partial r_0}{\partial \sigma_0} = \begin{cases} \frac{\partial r_0}{\partial \sigma_0} > 0 & \rho_0 > 1 \\ \frac{\partial r_0}{\partial \sigma_0} \leq 0 & \rho_0 \leq 1 \end{cases}. \tag{2.59}$$

Considering cell division as an example, the number of daughter cells is proportional to the size of the mother cell (of course, it is usually two). Hence, the fertility becomes

$$\phi(x^\diamond) = \frac{x^\diamond}{x}.$$

Substituting this into ρ_0 gives a value of one. In the real world, because reproduction involves thermodynamic or material losses, at least, ρ_0 must be less than one. Consequently, individual stochasticity can influence such semelparity negatively.

Now, for several demographic indices, let us focus on the mathematical structure of the characteristic function. In comparison with Eq. (2.51), the survivorship until the mature age is equivalent to a probability for reaching the mature size with survival:

$$\mathbb{E}_{X_0=x}\left[\exp\{-\mu_0 a^\diamond\}\right] = \left(\frac{x}{x^\diamond}\right)^{\rho_+}\Big|_{r=0}.$$

For the persistence of the species, they have to produce one or more offspring that survives until the mature age. Therefore,

$$1 \leq \phi(x^\diamond)\left(\frac{x}{x^\diamond}\right)^{\rho_+}\Big|_{r=0}.$$

Because the characteristic function is monotonically decreasing in r, it must exceed one at $r = 0$ for taking a positive r_0. According to the definition of the cumulant-generating function Θ_r (see Eq. (2.43)), it gives

$$\Theta_r = \ln\{\psi_r\} = \ln\{\phi(x^\diamond)\} - \rho_+ \ln\left\{\frac{x^\diamond}{x}\right\}.$$

As the first-order cumulant represents the mean mature age, the first derivative of the function above becomes

$$\langle a^{\diamond}\rangle^{(1)} = -\frac{d}{dr}\Theta_r\bigg|_{r=0} = \frac{\ln\left\{\frac{x^{\diamond}}{x}\right\}}{\sqrt{\left(g_0 - \frac{\sigma_0^2}{2}\right)^2 + \frac{\sigma_0^2\mu_0}{2}}}. \tag{2.60}$$

Unlike the deterministic size growth model $\sigma_0 = 0$, the mean mature age is linked to mortality. Higher mortality reduces the mature age because individuals with slow growth are at high risk of accidental death. In deterministic growth, all the individuals have identical growth rates; hence, this never occurs. Meanwhile, the life expectancy at birth is computed as

$$\begin{aligned} e_0 &= \int_0^{\infty} da\, \mathbb{E}_{X_0=x}\left[\exp\left\{-\int_0^a d\tau\, \mu_s(X_\tau)\right\}\right] \\ &= \mathbb{E}_{X_0=x}\left[\int_0^{a^{\diamond}} da\, \exp\{-\mu_0 a\}\right] \\ &= \frac{1}{\mu_0}\mathbb{E}_{X_0=x}\left[1-\exp\left\{-\mu_0 a^{\diamond}\right\}\right] \\ &= \frac{1}{\mu_0}\left(1-\left(\frac{x}{x^{\diamond}}\right)^{\rho_+}\right)\bigg|_{r=0}. \end{aligned} \tag{2.61}$$

Compared with the mean mature age, the life expectancy at birth is determined by the probability of reaching the mature age in addition to mortality. The statistics for the mature age consider individuals who succeed in reproduction, whereas the latter is subject to all individuals. If the growth is deterministic, $\sigma_0 = 0$, and individuals never die until the mature age, $\mu_0 = 0$, both indices are equivalent:

$$a^{\diamond} = e_0 = \frac{1}{g_0}\ln\left\{\frac{x^{\diamond}}{x}\right\}.$$

Determining the reproductive age distribution that gives the indices concerning maturity and longevity is essential for understanding semelparity. As we already know that the characteristic function constitutes the Laplace transform of the distribution, it is derived from its inverse transform.

$$\begin{aligned} \mathcal{R}_s(a) &= \lim_{\omega\uparrow\infty}\frac{1}{2\pi\mathrm{i}}\int_{c-\mathrm{i}\omega}^{c+\mathrm{i}\omega} dr\, \exp\{ra\}\frac{\psi_r(x)}{\psi_0(x)} \\ &= \frac{\phi(x^{\diamond})}{\psi_0(x)}\frac{\ln\left\{\frac{x^{\diamond}}{x}\right\}}{\sqrt{2\pi\sigma_0^2 a^3}}\exp\left\{\frac{-\left[\ln\left\{\frac{x^{\diamond}}{x}\right\}-\left(g_0-\frac{\sigma_0^2}{2}\right)a\right]^2}{2\sigma_0^2 a} - \mu_0 a\right\}. \end{aligned} \tag{2.62}$$

Let us prove the relation above. As a preparation, we present the following formula Oizumi et al. (2016):

$$\int_0^\infty da\, \frac{1}{\sqrt{a^3}} \exp\left\{-\left(\sqrt{c_1}\sqrt{a} - \frac{\sqrt{c_2}}{\sqrt{a}}\right)^2\right\} = \frac{\sqrt{\pi}}{\sqrt{c_2}} \quad c_1, c_2 > 0. \tag{2.63}$$

Because

$$\begin{aligned}
&\int_0^\infty da\, \frac{1}{\sqrt{a^3}} \exp\left\{-\left(\sqrt{c_1}\sqrt{a} - \frac{\sqrt{c_2}}{\sqrt{a}}\right)^2\right\} \\
&\quad = 2\int_0^\infty d\tau\, \frac{1}{\tau^2} \exp\left\{-\left(\sqrt{c_1}\tau - \frac{\sqrt{c_2}}{\tau}\right)^2\right\} \quad \sqrt{a} \to \tau,
\end{aligned}$$

it suffices to prove that

$$\int_0^\infty d\tau\, \frac{1}{\tau^2} \exp\left\{-\left(\sqrt{c_1}\tau - \frac{\sqrt{c_2}}{\tau}\right)^2\right\} = \frac{\sqrt{\pi}}{2\sqrt{c_2}}. \tag{2.64}$$

By changing the variable again, $\tau = \sqrt{\frac{c_2}{c_1}\frac{1}{s}}$, we can see that

$$\left(\sqrt{c_1}\tau - \frac{\sqrt{c_2}}{\tau}\right)^2 = \left(\sqrt{c_1}s - \frac{\sqrt{c_2}}{s}\right)^2,$$

and

$$\int_0^\infty d\tau\, \frac{1}{\tau^2} \exp\left\{-\left(\sqrt{c_1}\tau - \frac{\sqrt{c_2}}{\tau}\right)^2\right\} = \int_0^\infty ds\, \sqrt{\frac{c_1}{c_2}} \exp\left\{-\left(\sqrt{c_1}s - \frac{\sqrt{c_2}}{s}\right)^2\right\}. \tag{2.65}$$

From the above, the following holds:

$$\begin{aligned}
&\int_0^\infty d\tau\, \frac{1}{\tau^2} \exp\left\{-\left(\sqrt{c_1}\tau - \frac{\sqrt{c_2}}{\tau}\right)^2\right\} + \int_0^\infty d\tau\, \sqrt{\frac{c_1}{c_2}} \exp\left\{-\left(\sqrt{c_1}\tau - \frac{\sqrt{c_2}}{\tau}\right)^2\right\} \\
&= \int_0^\infty d\tau\, \left(\frac{1}{\tau^2} + \sqrt{\frac{c_1}{c_2}}\right) \exp\left\{-\left(\sqrt{c_1}\tau - \frac{\sqrt{c_2}}{\tau}\right)^2\right\} \\
&= \frac{1}{\sqrt{c_2}} \int_0^\infty d\tau\, \left[\frac{d}{d\tau}\left(\sqrt{c_1}\tau - \frac{\sqrt{c_2}}{\tau}\right)\right] \exp\left\{-\left(\sqrt{c_1}\tau - \frac{\sqrt{c_2}}{\tau}\right)^2\right\} \\
&= \frac{1}{\sqrt{c_2}} \int_{-\infty}^\infty d\theta\, \exp\left\{-\theta^2\right\} = \sqrt{\frac{\pi}{c_2}}, \quad \sqrt{c_1}\tau - \frac{\sqrt{c_2}}{\tau} \to \theta.
\end{aligned}$$

From this result and Eq. (2.65), Eq. (2.64) is verified. Multiplying Eq. (2.63) by

$$\frac{\ln\left\{\frac{x^\diamond}{x}\right\}}{\sqrt{2\pi\sigma_0^2}}\exp\left\{-c_3-2\sqrt{c_1c_2}\right\},\quad c_3\in\mathbb{R}$$

gives

$$\int_0^\infty da\,\frac{\ln\left\{\frac{x^\diamond}{x}\right\}}{\sqrt{2\pi\sigma_0^2a^3}}\exp\left\{-c_1a-\frac{c_2}{a}-c_3\right\}=\frac{\ln\left\{\frac{x^\diamond}{x}\right\}}{\sqrt{2\sigma_0^2c_2}}\exp\left\{-c_3-2\sqrt{c_1c_2}\right\}. \tag{2.66}$$

We rewrite each constant as follows:

$$c_1=\frac{\left(g_0-\frac{\sigma_0^2}{2}\right)^2}{2\sigma_0^2}+\mu_0+r,\quad c_2=\frac{\left(\ln\left\{\frac{x^\diamond}{x}\right\}\right)^2}{2\sigma_0^2},\quad c_3=\frac{\ln\left\{\frac{x^\diamond}{x}\right\}\left(g_0-\frac{\sigma_0^2}{2}\right)}{\sigma_0^2}.$$

Substituting these into Eq. (2.66), we have

$$\int_0^\infty da\,\frac{\ln\left\{\frac{x^\diamond}{x}\right\}}{\sqrt{2\pi\sigma_0^2a^3}}\exp\left\{\frac{-\left[\ln\left\{\frac{x^\diamond}{x}\right\}-\left(g_0-\frac{\sigma_0^2}{2}\right)a\right]^2}{2\sigma_0^2a}-(\mu_0+r)\,a\right\}=\left(\frac{x}{x^\diamond}\right)^{\rho_+}, \tag{2.67}$$

because $\frac{\ln\left\{\frac{x^\diamond}{x}\right\}}{\sqrt{2\sigma_0^2c_2}}=1$ and

$$-c_3-2\sqrt{c_1c_2}=-\rho_+\ln\left\{\frac{x^\diamond}{x}\right\}.$$

From Eq. (2.57), the mature age distribution corresponds with

$$\mathcal{Q}_s\left(0,x\to a,x^\diamond\right)=\frac{\ln\left\{\frac{x^\diamond}{x}\right\}}{\sqrt{2\pi\sigma_0^2a^3}}\exp\left\{\frac{-\left[\ln\left\{\frac{x^\diamond}{x}\right\}-\left(g_0-\frac{\sigma_0^2}{2}\right)a\right]^2}{2\sigma_0^2a}-\mu_0a\right\}. \tag{2.68}$$

The integrand in Eq. (2.62) is computed as

$$\frac{\psi_r(x)}{\psi_0(x)}=\left(\frac{x}{x^\diamond}\right)^{\rho_+(r)-\rho_+(0)}.$$

Hence, we get

$$\mathcal{R}_s(a) = \left(\frac{x^\diamond}{x}\right)^{\rho_+}\Big|_{r=0} \frac{\ln\left\{\frac{x^\diamond}{x}\right\}}{\sqrt{2\pi\sigma_0^2 a^3}} \exp\left\{\frac{-\left[\ln\left\{\frac{x^\diamond}{x}\right\} - \left(g_0 - \frac{\sigma_0^2}{2}\right)a\right]^2}{2\sigma_0^2 a} - \mu_0 a\right\}.$$

This completes the proof.

We have analyzed the statistics that link the life history and population dynamics of semelparous organisms without the structured population model. Setting the semelparous total newborns to β_t^s, the mature age density and the fertility function give the renewal process as follows:

$$\beta_t^s(x) = \int_t^\infty da \int_0^{x^\diamond} dy\, p(a-t, y)\, \mathcal{Q}_s\left(a-t, y \to a, x^\diamond\right) \phi\left(x^\diamond\right)$$
$$+ \int_0^t da\, \beta_{t-a}^s(x)\, \mathcal{Q}_s\left(0, x \to a, x^\diamond\right) \phi\left(x^\diamond\right). \tag{2.69}$$

The age and size transition process occurs within the mature age only, which does not concern reproduction. If $P_t^s(a, y)$ denotes the semelparous cohort density, it should satisfy

$$P_t^s(a, 0) = P_t^s\left(a, x^\diamond\right) = 0,$$

on

$$\left[\frac{\partial}{\partial t} + \frac{\partial}{\partial a}\right] P_t^s(a, 0) = -H_s(y)\, P_t^s(a, y) \tag{2.70}$$

$$-H_s(y)\, P_t^s(a, y) := \left[-\mu_0 - \frac{\partial}{\partial y} g_0 y + \frac{1}{2}\frac{\partial^2}{\partial y^2}\sigma_0^2 y^2\right] P_t^s(a, y). \tag{2.71}$$

Here, $y = 0$ implies a fixed point where the size transition stops. Because the SDE model has no exploded solution[15], the size never reaches zero at a finite time (Cf. Eq. (1.4)). Hence, this is the natural boundary. Conversely, $y = x^\diamond$ is reachable at a finite time. However, individuals reaching the mature size reproduce and die simultaneously. Hence, the condition is interpretable as an absorbing boundary. As we already know that the population density P_t^s consists of the state transition kernel, let us consider the kernel under the boundary condition.

---

[15] It implies that the solution diverges at a finite time, such as

$$\lim_{t \uparrow 1} \frac{1}{1-t} = \infty.$$

$$\frac{\partial}{\partial a} K^{s}(0, x \to a, y) = -H_{s}(y) K^{s}(0, x \to a, y)$$
$$K^{s}(0, x \to 0, y) = \delta(x - y)$$
$$K^{s}(0, x \to a, 0) = K^{s}\left(0, x \to a, x^{\diamond}\right) = 0.$$

To simplify this equation, we apply a variable transformation such that

$$z = \ln\{y\}, \quad K^{s}(0, x \to a, z) = \exp\{-\mu_0 a\} f(a, z).$$

Then, the transition kernel consisting of the new variable follows

$$\frac{\partial}{\partial a} f(a, z) = -\left(g_0 - \frac{\sigma^2}{2}\right)\frac{\partial}{\partial z} f(a, z) + \frac{\sigma^2}{2}\frac{\partial^2}{\partial z^2} f(a, z) \tag{2.72}$$

$$f(0, z) = \delta(x - \exp\{z\}) \tag{2.73}$$

$$\lim_{z \downarrow -\infty} f(a, z) = \lim_{z \uparrow z^{\circ}} f(a, z) = 0, \quad z^{\diamond} := \ln\left\{x^{\diamond}\right\}. \tag{2.74}$$

The solution for the above equation is given by

$$f(a, z) = \frac{1}{\sqrt{2\pi\sigma_0^2 a}}\left[\exp\left\{-\frac{\left(z - z_0 - \left(g_0 - \frac{\sigma^2}{2}\right)a\right)^2}{2\sigma_0^2 a}\right\}\right.$$
$$\left. - \exp\left\{-\frac{2\left(g_0 - \frac{\sigma^2}{2}\right)\left(z_0 - z^{\diamond}\right)}{\sigma_0^2} - \frac{\left(z + z_0 - 2z^{\diamond} - \left(g_0 - \frac{\sigma^2}{2}a\right)a\right)^2}{2\sigma_0^2 a}\right\}\right],$$

where $z_0 := \ln\{x\}$ Goel et al. (1974). With the original parameters, the following relationship holds:

$$\int_0^{x^{\diamond}} dy\, K^{s}(0, x \to a, y) = \int_{-\infty}^{z^{\diamond}} dz\, K^{s}(0, x \to a, z)$$
$$= \int_0^{x^{\diamond}} \frac{dy}{y} \exp\{-\mu_0 a\} f(a, y).$$

Therefore, we obtain

$$K^{s}(0, x \to a, y) = \frac{1}{y\sqrt{2\pi\sigma_0^2 a}}\left[\exp\left\{-\frac{\left(\ln\left\{\frac{y}{x}\right\} - \left(g_0 - \frac{\sigma^2}{2}\right)a\right)^2}{2\sigma_0^2 a} - \mu_0 a\right\}\right.$$
$$\left. - \exp\left\{\frac{2\left(g_0 - \frac{\sigma^2}{2}\right)\ln\left\{\frac{x^{\diamond}}{x}\right\}}{\sigma_0^2} - \frac{\left(\ln\{yx\} - 2\ln\left\{x^{\diamond}\right\} - \left(g_0 - \frac{\sigma^2}{2}\right)a\right)^2}{2\sigma_0^2 a} - \mu_0 a\right\}\right]. \tag{2.75}$$

In addition, we get the stable age and multi-state distribution, which is the eigenfunction w_0^{s} corresponding to the dominant characteristic root (see Eq. (2.58)):

$$w_0^{\mathrm{s}}(a, y) = \exp\{-r_0 a\} K^{\mathrm{s}}(0, x \to a, y).$$

From the results above, we can explicitly formulate the population dynamics of semelparity with the stochastic size growth.

$$P_t^{\mathrm{s}}(a, y) = \begin{cases} \int_0^{x^\diamond} d\xi \; p(a-t, \xi) K^{\mathrm{s}}(a-t, \xi \to a, y) & t-a \le 0 \\ \beta_t^{\mathrm{s}}(x) K^{\mathrm{s}}(0, x \to a, y) & t-a > 0 \end{cases} \Leftarrow \text{Eq. (36)}$$

$$\beta_t^{\mathrm{s}}(x) = \int_t^{\infty} da \int_0^{x^\diamond} dy \; p(a-t, y) \mathcal{Q}_{\mathrm{s}}\left(a-t, y \to a, x^\diamond\right) \phi\left(x^\diamond\right) + \int_0^t da \; \beta_{t-a}^{\mathrm{s}}(x) \mathcal{Q}_{\mathrm{s}}\left(0, x \to a, x^\diamond\right) \phi\left(x^\diamond\right) \Leftarrow \text{Eq. (209)}.$$

Thus, this model is reasonable for understanding our computation and it shows that our framework can deal with semelparous life histories. Moreover, with the state growth model, fertility, neonatal state, and mortality, we can derive various demographic indices regarding vicissitudes of population and life from the characteristic function without the structured population model. These features characterize species and provide clues for their evolution. Biologists believe that optimized life histories under natural selection determine the *mean lifespan*, *intrinsic rate of natural increase*, *BRN*, and *reproductive value*.

According to an early interpretation of the evolutionary theory, the fittest species maximizes its population growth rate. Thus, the adaptive life history maximizes the intrinsic rate of natural increase. Because the semelparous mathematical model shows that the variation of size growth reduces the population growth (see Eq. (2.59)), there is a possibility that semelparous species evolve to avoid individual stochasticity. Furthermore, under high mortality, there should be precocity in comparison with deterministic life histories (see Eq. (2.60)). Thus, we can identify various features of individual stochasticity. These traits arise from the statistics of sample paths that do not occur in ODE models. Consequently, individual stochasticity providing diversity within a population indicates a close relationship with population growth.

References

Preston, S., Coale, A. J., Trussell, J., & Weinstein, M. (1980). Estimating the completeness of reporting of adult deaths in populations that are approximately stable. *Population Index*, 179–202.

Rogers, A., & Ledent, J. (1976). Increment-decrement life tables: a comment. *Demography*, 287–290.

Lv, Q., & Pitchford, J. W. (2007). Stochastic von bertalanffy models, with applications to fish recruitment. *Journal of Theoretical Biology, 244*(4), 640–655.

Banks, H. T., Davis, J. L., Ernstberger, S. L., Hu, S., Artimovich, E., Dhar, A., & Browdy, C. (2009). A comparison of probabilistic and stochastic formulations in modelling growth uncertainty and variability. *Journal of Biological Dynamics, 3*(2), 130–148.

Langlais, M. (1988). Large time behavior in a nonlinear age-dependent population dynamics problem with spatial diffusion. *Journal of Mathematical Biology, 26*(3), 319–346.

Metz, J., & Diekmann, O. (1986). *The dynamics of physiologically structured populations* (Vol. 68). Springer.

Iannelli, M. (1995). Mathematical theory of age-structured population dynamics. *Giardini Editori e Stampatori in Pisa.*

Inaba, H. (2017). Age-structured population dynamics in demography and epidemiology.

Øksendal, B. (2003). *Stochastic differential equations: an introduction with applications*. Berlin: Springer.

Feller, W. (1941). On the integral equation of renewal theory. *The Annals of Mathematical Statistics*, 243–267.

Feynman, R., Deutsch, A., Eckart, C., Cowen, R., Dieke, G., & Society, A. P. (1948). *Space-time approach to non-relativistic quantum mechanics*. American Physical Society.

Kac, M. (1949). On distributions of certain wiener functionals. *Transactions of the American Mathematical Society, 65*(1), 1–13.

Fisher, R. A. (1930). *The genetical theory of natural selection*. Oxford: Oxford University Press.

Oizumi, R., & Takada, T. (2013). Optimal life schedule with stochastic growth in age-size structured models: theory and an application. *Journal of Theoretical Biology, 323*, 76–89.

Oizumi, R., Kuniya, T., & Enatsu, Y. (2016). Reconsideration of r/k selection theory using stochastic control theory and nonlinear structured population models. *PloS one, 11*(6), e0157715.

Goel, N., & Richter-Dyn, N. (1974). *Stochastic models in biology*. London: Academic Press.

Chapter 3
Non-linear Structured Population Models

The previous chapters mainly treated linear structured population models. However, there are just three simple asymptotic behaviors, e.g., the population dynamics is monotonically increasing, $r_0 > 0$, decreasing, $r_0 < 0$, or constant, $r_0 = 0$. In nature, most organisms persist via complicated population dynamics established over centuries. Biologists believe that density effects as well as intra-specific and inter-specific competitions are significant phenomena that complicate the population dynamics. These phenomena appear in mathematical models for biology (including human demography) as non-linear effects. Such non-linearity is a central subject in both mathematical demography and biology. This chapter deals with the analysis of non-linear effects on individual life histories and population dynamics.

3.1 r/K Selection Theory

The population will not be able to conserve exponential growth as with the Malthus equation and linear structured population models, because resources in the world are limited. Many readers will probably agree with this, and the limit also matches with physical restrictions. However, it is also a fact that the fittest species multiply exponentially in a short time. For example, mice and locusts often exhibit a massive outbreak under an abundance of resources. Human beings have also shown enormous growth since the discovery of fossil fuels.

In contrast to these species, we have never heard of a considerable outbreak of elephants or whales. If all the fittest species maximize their Malthus parameters or the intrinsic rate of natural increase, it is not easy to understand the life history of the latter species. Large mammals such as elephants have a late maturity age and a long gestation period, whereas mice, which have precocity and prolificacy, should take their place. Nevertheless, biologists believe that such natural selection does not

R. Oizumi, *Population Dynamics Based on Individual Stochasticity*,
Population Studies of Japan,
https://doi.org/10.1007/978-981-19-3548-0_3

exist. They believe that there are various fittest species with optimization for each environment.

From an evolutionary point of view, knowing what the fittest species optimizes is critical for understanding its life history. Here, we introduce a classical hypothesis or traditional interpretation that is called the r/K selection theory. This hypothesis originated from a textbook by MacArthur and Wilson MacArthur and Wilson (1967). It describes the evolution of life histories with different tendencies of optimization for each degree of the density effect. They correctly assumed that the population dynamics of organisms follows the logistic equation

$$\frac{dN_t}{dt} = rN_t\left(1 - \frac{N_t}{K}\right),$$

where N_t and K denote the total population density and the carrying capacity[1], respectively. When the population density is low, such as at the beginning of primary succession $N_t/K \approx 0$, the theory argues that species with the maximum r dominate their habitat.

Conversely, they stated that species that maximize their carrying capacity have greater advantages than otherwise when the population size is around K. According to their description, under the saturated population, ensuring that one descendant survives is more adaptive than prolific birth for high population growth. Therefore, the authors distinguished the r-strategy for maximizing the population growth from the K-strategy for maximizing the carrying capacity.

Subsequently, Pianka classified life strategies based on several vital rates of species Pianka (1976). In summary, as typical traits of the r-strategy, species have a precocial, prolific, and short lifespan, whereas the K-strategy involves late maturity, low fertility, and longevity. This classification caused controversy among biologists for some decades. Biologists have provided many positive and negative pieces of evidence to support the conjecture; however, the discussion faded without solving the problems until it resurfaced in recent years Pianka (2002).

There are several logical gaps and exceptions in Pianka's classification. Notable points are related to the K-strategy, e.g., how could late maturity and low fertility maximize the carrying capacity. Accordingly, biologists have still not been able to determine the appropriate classification for the K-strategy.

This theory, based on the logistic equation, seems to suggest that species maximize one of the two independent parameters as required. The setting r can be associated with the dominant characteristic root of structured population models; hence, it is closely related to the life history. In contrast with r, the carrying capacity cannot be derived from the vital rates of each life history because it originates from the density effect. To deal with this problem, we must consider a structured population model with the density effect. We believe that the treatment of the density effect is critical for determining the K-strategy. Next, we address this problem.

[1] It represents the capacity of the population that the environment feeds.

3.2 Non-linear Multi-state Population Model

Density effects are incorporated in many phenomena such as intra-specific and inter-specific competition as well as infectious diseases. Reproduction requires a fixed population density because individuals of the opposite sex must encounter each other at an appropriate period, which is called the Allee effect Allee and Bowen (1932). Meanwhile, overpopulation involves risks such as death by infectious diseases, war, carnivorism, and famines. Conversely, in contemporary society, communication of individuals promotes the development of science, technology, and medicine. Hence, the mortality tends to decrease. Moreover, the density effect influences not only birth and death but also state transitions, such as migration and size growth, owing to economic factors or intra-specific competition.

In general, the density effect depends on the temporal population structure. Each age and size has a different impact on each life history. For example, individuals with a large body size consume many resources, whereas small juveniles and elderly individuals may not make many contributions to intra-specific competition because of their low consumption. Considering various cases of interaction among individuals, let the following function be a vector Γ_t representing the density effect.

$$\Gamma_t = (\Gamma_t(\ell))_{1 \le \ell \le m} := (\langle \gamma_\ell, P_t \rangle)_{1 \le \ell \le m}, \quad \gamma_\ell(a, y) \ge 0, \quad \forall \ell. \tag{3.1}$$

In general, the vector affects all the vital parameters, such as state transition, fertility, and mortality. Accordingly, Eqs. (2.1) and (2.2) and the fertility are rewritten as follows:

$$\begin{aligned} dX_a^j &= g_j(a, X_a, \Gamma_t)\, da + \sum_{k=1}^{n} \sigma_{jk}(a, X_a, \Gamma_t)\, dB_a^k, \quad X_s^j = x^j, \\ &\sigma_{jk}(a, X_a) \ge 0 \quad j = 1, 2, \cdots, d, \end{aligned} \tag{3.2}$$

$$\mu(a, x, \Gamma_t) > 0, \quad \int_0^\alpha da\, \mu(a, x, \Gamma_t) = +\infty \quad \text{for all } x \in A, \tag{3.3}$$

and

$$F(a, x, \Gamma_t) \ge 0. \tag{3.4}$$

The population dynamics that these ingredients of life histories (Eqs. (3.2), (3.3), and (3.4)) constitute is given by a derivation that is identical to that of the linear model Eq. (2.8) as follows:

$$\left[\frac{\partial}{\partial t}+\frac{\partial}{\partial a}\right] P_t(a,y) = -H(a,y,\Gamma_t)\, P_t(a,y) \tag{3.5}$$

$$\begin{aligned} H(a,y,\Gamma_t)\, P_t(a,y) := {} & \mu(a,y,\Gamma_t)\, P_t(a,y) \\ & + \sum_j \frac{\partial}{\partial y^j}\left(g_j(a,y,\Gamma_t)\, P_t(a,y)\right) \\ & - \frac{1}{2}\sum_{j=1}^{d}\sum_{j'=1}^{d} \frac{\partial^2}{\partial y^j \partial y^{j'}}\left(S_{jj'}(a,y,\Gamma_t)\, P_t(a,y)\right). \end{aligned} \tag{3.6}$$

As with the reproduction process in the linear model (see Eq. (2.9)), the newborn dynamics follows the integral equation

$$\beta_t = \langle F(\Gamma_t), P_t \rangle = \int_0^\infty da \int_A dy\; F(a,y,\Gamma_t)\, P_t(a,y), \quad F(a,y,\Gamma_t) \geq 0. \tag{3.7}$$

Similarly, we reuse the assumption concerning the neonatal state distribution Eq. (2.10).

3.3 Stability of Stationary Population

The density effects are thought to not only complicate the population dynamics but also suppress the population growth. In general, it is difficult for non-linear population dynamics to obtain the explicit mathematical form. Instead of obtaining the exact solution, mathematicians focus on the stability of a stationary population. Local stability analysis is a typical method for categorizing the population dynamics. The idea is that a robust stationary population reduces the first-order perturbation around itself.

3.3.1 *Stationary Population*

Let $P^\dagger(a,y) \geq 0$ be a stationary population of Eq. (3.5). Then, $P^\dagger(a,y)$ satisfies

$$\frac{\partial}{\partial a} P^\dagger(a,y) = -H\left(a,y,\Gamma^\dagger\right) P^\dagger(a,y) \tag{3.8}$$

$$P^\dagger(0,y) = \beta^\dagger \nu(y) \tag{3.9}$$

$$\Gamma^\dagger := \left(\left\langle \gamma_\ell, P^\dagger \right\rangle\right)_{1\leq \ell \leq m}. \tag{3.10}$$

$$\beta^\dagger := \left\langle F\left(\Gamma^\dagger\right), P^\dagger \right\rangle. \tag{3.11}$$

Equation (3.8) is the Fokker–Planck equation because $\Gamma^\dagger$ and $\beta^\dagger$ are constant. Hence, we obtain the following solution (see Eq. (2.5)):

$$P^\dagger(a, y) = \beta^\dagger \int_A dx\, \nu(x)\, K^\dagger(0, x \to a, y), \tag{3.12}$$

where $K^\dagger(\cdot)$ is the fundamental solution (cf. Eq. (2.3)) of

$$\frac{\partial}{\partial a} K^\dagger(a, y) = -H\left(a, y, \Gamma^\dagger\right) K^\dagger(a, y), \quad K^\dagger(0, y) = \delta^d(x - y). \tag{3.13}$$

Substituting Eq. (3.12) into an element of Eq. (3.1), we have

$$\Gamma^\dagger(\ell) = \beta^\dagger \int_0^\infty da \int_A dy \int_A dx\, \nu(x)\, K^\dagger(0, x \to a, y)\, \gamma_\ell(a, y). \tag{3.14}$$

Here, the relationship between the total stationary newborns $\beta^\dagger$ and each density effect $\Gamma^\dagger(\ell)$ is represented by

$$\beta^\dagger = \frac{\Gamma^\dagger(\ell)}{\int_0^\infty da \int_A dy \int_A dx\, \nu(x)\, K^\dagger(0, x \to a, y)\, \gamma_\ell(a, y)}, \quad \ell = 1, 2, \cdots m. \tag{3.15}$$

To specify $\Gamma^\dagger$, we derive the characteristic equation by substituting Eq. (3.12) into Eq. (3.11), and we get

$$1 = \int_0^\infty da \int_A dy \int_A dx\, \nu(x)\, K^\dagger(0, x \to a, y)\, F\left(a, y, \Gamma^\dagger\right). \tag{3.16}$$

Therefore, each $\Gamma^\dagger(\ell)$ must satisfy Eq. (3.16). Furthermore, note that each component $\Gamma^\dagger(\ell)$ of $\Gamma^\dagger$ is independent of the mass function $\gamma_\ell(\cdot)$ for the density effect. Thus, the stationary population is given by

$$P^\dagger(a, y) = \frac{\Gamma^\dagger(\ell) \int_A dx\, \nu(x)\, K^\dagger(0, x \to a, y)}{\int_0^\infty ds \int_A d\xi \int_A dx\, \nu(x)\, K^\dagger(0, x \to s, \xi)\, \gamma_\ell(s, \xi)}, \quad \ell = 1, 2, \cdots, m. \tag{3.17}$$

A feature of this framework is that the traits of the intra-specific competition among ages and states $\gamma_\ell(\cdot)$ do not affect the stationary population density because $\beta^\dagger$ is constant. Meanwhile, $\Gamma^\dagger$ is involved in the stationary population structure as the force of the density effect because it affects individual life histories. Although each $\gamma_\ell(\cdot)$ as an interaction function does not affect the life history, it plays a significant role in determining the size of the total stationary population.

3.3.2 Characteristic Equation for Local Stability

The local stability of the stationary population density is important information for the eventual population dynamics. If the stationary population is stable, species can recover their population to a steady state under disturbances. This topic deals with the characteristic equation and its roots that discriminate the stability.

Let $\delta P_t(a, y)$ be an infinitesimal difference from the stationary population density $P^\dagger(a, y)$.

$$\delta P_t(a, y) := P_t(a, y) - P^\dagger(a, y).$$

Then, each difference of Γ_t becomes

$$\delta\Gamma_t = (\delta\Gamma_t(\ell))_{1\le\ell\le m} := (\langle\gamma_\ell, \delta P_t\rangle)_{1\le\ell\le m}, \quad \gamma_\ell(a, y) \ge 0, \quad \forall\ell.$$

The neighborhood dynamics of the stationary population is given by the substitution of the following equations into Eqs. (3.5) and (3.7).

$$P_t(a, y) = P^\dagger(a, y) + \delta P_t(a, y), \quad \Gamma_t = \Gamma^\dagger + \delta\Gamma_t.$$

Hence, the renewal equation given by Eq. (3.7) becomes

$$\beta_t + \delta\beta_t = \left\langle F\left(\Gamma^\dagger + \delta\Gamma_t\right), \left(P^\dagger + \delta P_t\right)\right\rangle.$$

Let us consider sufficiently small $\delta P_t(\cdot)$ and $\delta\Gamma_t$ such that the first-order approximation holds. Each ingredient constituting Eqs. (3.5) and (3.7) is approximated by

$$\left[\frac{\partial}{\partial t} + \frac{\partial}{\partial a}\right]\delta P_t(a, y) = -H\left(a, y, \Gamma^\dagger\right)\delta P_t(a, y) - \sum_{\ell=1}^{m}\delta\Gamma_t(\ell)\,\theta_\ell\left(a, y, \Gamma^\dagger\right) \tag{3.18}$$

$$\theta_\ell(a, y) := \frac{\partial}{\partial\Gamma^\ell}H(a, y, \Gamma)\Big|_{\Gamma=\Gamma^\dagger} P^\dagger(a, y)$$
$$\delta P_0(a, y) = p(a, y) - P^\dagger(a, y)$$
$$\delta P_t(0, y) = \delta\beta_t\nu(y)$$
$$\delta\beta_t := \left\langle F\left(\Gamma^\dagger\right), \delta P_t\right\rangle + \sum_{\ell=1}^{m}\delta\Gamma_t(\ell)\left\langle F'_\ell\left(\Gamma^\dagger\right), P^\dagger\right\rangle \tag{3.19}$$
$$F'_\ell\left(a, y, \Gamma^\dagger\right) := \frac{\partial}{\partial\Gamma^\ell}F(a, y, \Gamma)\Big|_{\Gamma=\Gamma^\dagger}.$$

Then, the second-order and higher-order terms are assumed to be zero, i.e.,

$$\begin{aligned}
&\delta P_t(a, y)^2 = \delta P_t(a, y)\,\delta\Gamma_t(\ell) = \delta\Gamma_t(\ell)^2 = \delta\beta_t^2 = 0, \\
&\delta\Gamma_t(\ell)\, H\left(a, y, \Gamma^\dagger\right) \delta P_t(a, y) = 0
\end{aligned} \qquad \ell = 1, \cdots, m. \tag{3.20}$$

This type of linear equation given by Eq. (3.18) has the exponential solution

$$\delta P_t(a, y) \equiv C_\lambda \exp\{\lambda t\}\, \mathrm{w}_\lambda(a, y), \quad \lambda \in \mathbb{C}.$$

Thus, by substituting it into Eqs. (3.18) and (3.19), we obtain the characteristic equation discriminating the stability of the stationary population. Then, the unknown function $\mathrm{w}_\lambda(a, y)$ satisfies

$$\left[\lambda + \frac{\partial}{\partial a}\right] \mathrm{w}_\lambda(a, y) = -H\left(a, y, \Gamma^\dagger\right) \mathrm{w}_\lambda(a, y) - \sum_{\ell=1}^{m} \langle \gamma_\ell, \mathrm{w}_\lambda \rangle\, \theta_\ell\left(a, y, \Gamma^\dagger\right), \tag{3.21}$$

and

$$\mathrm{w}_\lambda(0, y) = \left[\left\langle F\left(\Gamma^\dagger\right), \mathrm{w}_\lambda \right\rangle + \sum_{\ell=1}^{m} \langle \gamma_\ell, \mathrm{w}_\lambda \rangle \left\langle F'_\ell\left(\Gamma^\dagger\right), P^\dagger \right\rangle\right] \nu(v). \tag{3.22}$$

Because the terms inside the parentheses $[\cdots]$ are constant with respect to y, the function at age zero can be separated by the constant and the initial state distribution.

$$\mathrm{w}_\lambda(0, y) = \mathrm{w}_\lambda(0)\, \nu(y), \quad \mathrm{w}_\lambda(0) \equiv \left\langle F\left(\Gamma^\dagger\right), \mathrm{w}_\lambda \right\rangle + \sum_{\ell=1}^{m} \langle \gamma_\ell, \mathrm{w}_\lambda \rangle \left\langle F'_\ell\left(\Gamma^\dagger\right), P^\dagger \right\rangle.$$

Here, we introduce a new operator $T(a; s)$ that yields

$$\begin{aligned}
&T_\lambda(a; s)\,\phi(s, y) := \int_A dx\, \phi(s, x) \exp\{-\lambda(a - s)\}\, K^\dagger(s, x \to a, y) \quad 0 \le s \le a \\
&\lim_{a \downarrow s} T_\lambda(a; s) = I,
\end{aligned}$$

where I represents an identity operator:

$$I\phi(a, y) = \phi(a, y)$$

From the variational constant methodology, we computed the formal solution of Eq. (3.21) as

$$\mathrm{w}_\lambda(a,y) = \mathrm{w}_\lambda(0)\,T_\lambda(a;0)\,\nu(y) - \sum_{\ell=1}^{m} \langle \gamma_\ell, \mathrm{w}_\lambda\rangle \int_0^a ds\, T_\lambda(a;s)\,\theta_\ell\left(s,y,\Gamma^\dagger\right). \tag{3.23}$$

Computing $\langle \gamma_\ell, \mathrm{w}_\lambda\rangle$ using the equation above, we obtain

$$\langle \gamma_\ell, \mathrm{w}_\lambda\rangle = \mathrm{w}_\lambda(0)\left\langle \gamma_\ell, W_\lambda^\dagger\right\rangle - \sum_{k=1}^{m} \left\langle \gamma_\ell, \mathcal{T}_{\lambda,k}\right\rangle \langle \gamma_\ell, \mathrm{w}_\lambda\rangle \tag{3.24}$$

$$W_\lambda^\dagger(a,y) = \exp\{-\lambda a\} \int_A dx\, \nu(x)\, K^\dagger(0, x \to a, y)$$

$$\mathcal{T}_{\lambda,k}(a,y) = \int_0^a ds\, \exp\{-\lambda(a-s)\} \int_A dx\, \theta_k\left(s,x,\Gamma^\dagger\right) K^\dagger(s, x \to a, y).$$

This equation is solvable such that

$$\left(\langle \gamma_\ell, \mathrm{w}_\lambda\rangle\right)_{1\le\ell\le m} = \mathrm{w}_\lambda(0)\left(\mathbf{I} + \mathbf{T}_\lambda\right)^{-1} \mathbf{w}_\lambda^\dagger \tag{3.25}$$

$$\mathbf{I} : m \times m - - \text{Identity matrix}$$

$$\mathbf{T}_\lambda := \left(\left\langle \gamma_\ell, \mathcal{T}_{\lambda,k}\right\rangle\right)_{1\le k,\ell\le m}$$

$$\mathbf{w}_\lambda^\dagger := \left(\left\langle \gamma_\ell, W_\lambda^\dagger\right\rangle\right)_{1\le\ell\le m}.$$

From this result, the solution of Eq. (3.23) becomes

$$\mathrm{w}_\lambda(a,y) = \mathrm{w}_\lambda(0)\, W_\lambda^\dagger(a,y) - \mathrm{w}_\lambda(0)\,\mathbf{j}_\lambda^\top \left(\mathbf{I} + \mathbf{T}_\lambda\right)^{-1} \mathbf{w}^\dagger \tag{3.26}$$

$$\mathbf{j}_\lambda := \left(\mathcal{T}_{\lambda,k}(a,y)\right)_{1\le k\le m}.$$

Substituting Eqs. (3.25) and (3.26) into $\mathrm{w}_\lambda(0)$, we have

$$\mathrm{w}_\lambda(0) = \mathrm{w}_\lambda(0)\left[\psi_\lambda\left(\Gamma^\dagger\right) - \mathbf{f}_\lambda^\top \left(\mathbf{I} + \mathbf{T}_\lambda\right)^{-1} \mathbf{w}_\lambda^\dagger\right]. \tag{3.27}$$

$$\psi_\lambda\left(\Gamma^\dagger\right) := \left\langle F\left(\Gamma^\dagger\right), W_\lambda^\dagger\right\rangle \tag{3.28}$$

$$\mathbf{f}_\lambda := \left(\left\langle F\left(\Gamma^\dagger\right), \mathcal{T}_{\lambda,\ell}\right\rangle - \left\langle F_\ell'\left(\Gamma^\dagger\right), P^\dagger\right\rangle\right)_{1\le\ell\le m}.$$

Therefore, the characteristic roots $\lambda_k \in \mathbb{C}$ generating the exponential solution satisfy

$$1 = \psi_\lambda\left(\Gamma^\dagger\right) - \mathbf{f}_\lambda^\top \left(\mathbf{I} + \mathbf{T}_\lambda\right)^{-1} \mathbf{w}_\lambda^\dagger. \tag{3.29}$$

Let $\Lambda \subset \mathbb{C}$ be the set of characteristic roots given by the above characteristic equation. Then, the asymptotic behavior of the infinitesimal difference $\delta P_t(a,y)$ is given by the linear combination of the corresponding exponential solutions:

$$\delta P_t(a,y) = \sum_{k\in N(\Lambda)} C_k \exp\{\lambda_k t\}\, \mathrm{w}_k(a,y),$$

where $N(\Lambda)$ denotes the set of characteristic roots. If all the characteristic roots $\lambda_k \in \Lambda$ have negative real parts, then the stationary population is asymptotically stable Oizumi et al. (2016). Notably, the characteristic equation Eq. (3.29) coincides with the Euler–Lotka equation of our framework, i.e., Eq. (2.26), in $\Gamma^\dagger = 0$ because $\mathbf{f}_\lambda = 0$.

3.4 Concrete Example

In general, it is technically complex to analyze the mathematical properties of non-linear PDEs. In addition, it may not be easy to prove their local stability in many cases, even if the method presented in the previous section of this chapter is used. Nevertheless, a solvable example can elucidate life histories under non-linear population dynamics. This section deals with the semelparous life history, which we have already treated, by extending it to include intra-specific competition.

3.4.1 *Resource Acquisition Competition Model and Stationary Population*

Intra-specific competition for foraging reduces the resource allocation of individuals. It influences growth, fertility, and mortality rates. The depletion of nutrient resources leads to slow growth, late maturity, and a high risk of death for organisms. For simplicity, let us consider a case in which an increase in the population density reduces the mean size growth rate of individuals.

Suppose that the following density effect is used to rewrite the mean growth term of Eq. (2.46).

$$dX_a = (g_0 - \Gamma_t)\, X_a da + \sigma_0 X_a dB_a, \quad X_0 = x > 0. \tag{3.30}$$

Furthermore, $\gamma(a, y)$ is assumed to be constant.

$$\Gamma_t = \gamma_0 \int_0^\infty da \int_0^{x^\diamond} dy\ P_t^s(a, y), \quad \gamma(a, y) = \gamma_0 \equiv const. \tag{3.31}$$

Although this assumption is one of the simplest in biology, it is not realistic. Nevertheless, it does not influence the life histories of individuals because the solution of Eq. (3.16) gives the stationary state $\Gamma^\dagger$. The mathematical form $\gamma(a, y)$ determines the stationary population size and its local stability only (see Eq. (3.17) and the previous section of this chapter). As for mortality and fertility, we reuse the semelparity model mentioned previously, i.e., Eqs. (2.47) and (2.48).

Because the mean size growth rate decreases as the population increases, individuals will reach the mature size $x^\diamond$ at a late age and are at risk of accidental death. Then, an equilibrium between the size growth rate and the death rate will saturate the population. We refer to this model as the *resource acquisition competition* model. To examine this assumption, let us verify the existence of the stationary population.

As $\Gamma^\dagger$ is a scalar constant in this model, we can reuse many results in Section 2.5. Accordingly, we can replace g_0 to $g_0 - \Gamma^\dagger$ in Eq. (2.46) or substitute $\Gamma^\dagger$ into Γ_t of Eq. (3.30). $\Gamma^\dagger$ is derived from Eq. (3.16), whereas the stationary population should have $r_0 = 0$ in $\psi_{r_0} = 1$ (see Eq. (2.26)). Owing to the mathematical form of Eqs. (2.24), (3.16), and (3.28) under $r = \lambda = 0$,

$$\psi_r\Big|_{r=0} = \int_0^\infty da \int_0^{x^\diamond} dy \int_0^{x^\diamond} dx\, v(x)\, K^\dagger(0, x \to a, y)\, F\left(a, y, \Gamma^\dagger\right) = \psi_0\left(\Gamma^\dagger\right) = 1,$$

we get the characteristic function $\psi_0\left(x, \Gamma^\dagger\right)$ by solving the following equation (see Eq. (2.54)):

$$0 = -H_{\mathrm{s}}^*\left(y, \Gamma^\dagger\right) v(y) + F_{\mathrm{s}}(y), \tag{3.32}$$

where

$$H_{\mathrm{s}}^*\left(y, \Gamma^\dagger\right) = \mu_0 - \left(g_0 - \Gamma^\dagger\right) y \frac{d}{dy} - \frac{\sigma_0^2}{2} y^2 \frac{d^2}{dy^2}. \tag{3.33}$$

As $V(x) = \psi_0\left(x, \Gamma^\dagger\right)$ in this case, from Eq. (2.58), we get

$$0 = \mu_0 - \left(g_0 - \Gamma^\dagger\right) \rho_0 - \frac{\sigma_0^2}{2} \rho_0 (\rho_0 - 1), \quad \rho_0 = \frac{\ln\{\phi(x^\diamond)\}}{\ln\left\{\frac{x^\diamond}{x}\right\}}. \tag{3.34}$$

Therefore, the stationary density effect becomes

$$\Gamma^\dagger = g_0 - \frac{\sigma_0^2}{2}(1 - \rho_0) - \frac{\mu_0}{\rho_0}. \tag{3.35}$$

If $\Gamma^\dagger > 0$, the stationary population $N_{\mathrm{s}}^\dagger$ exists and it is

$$N_{\mathrm{s}}^\dagger := \int_0^\infty da \int_0^{x^\diamond} dy\, P^\dagger(a, y) = \frac{1}{\gamma_0}\Gamma^\dagger, \tag{3.36}$$

owing to Eq. (3.31). Thus, our framework links the size growth, fertility, and mortality to the stationary population size. For instance, Eq. (3.35) implies that the individual stochasticity σ_0 reduces the size of the population in $\rho_0 < 1$ (Fig. 3.1).

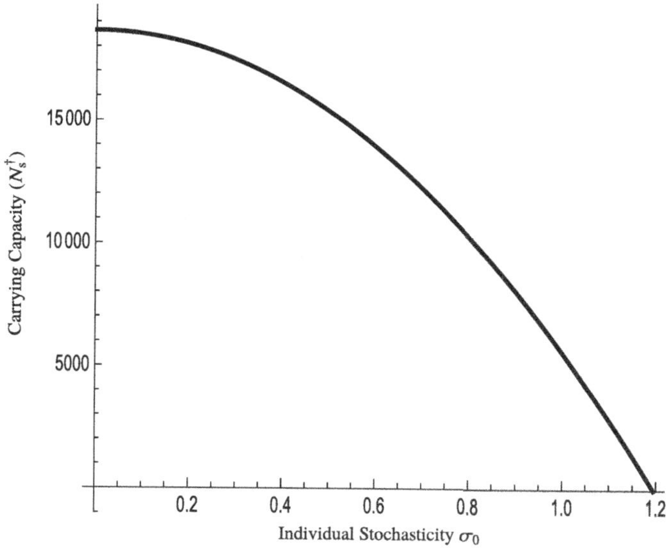

Fig. 3.1 Relationship between individual stochasticity and carrying capacity: This figure illustrates the effect of individual stochasticity on the carrying capacity. It indicates that diversity in the size growth rate decreases the carrying capacity of the environment. If the variability in the size growth rate is too high, the population will become extinct (the parameters are $g_0 = 0.2$, $x = 1$, $x^\diamond = 100$, $\phi(x^\diamond) = 30$, and $\gamma_0 = 1.0 \times 10^{-5}$)

3.4.2 *Local Stability of Stationary Population in Semelparous Species*

Now, let us consider the local stability of $N_s^\dagger$. The ingredients of the characteristic equation are given by

$$\psi_\lambda\left(x, \Gamma^\dagger\right) = \phi\left(x^\diamond\right)\left(\frac{x}{x^\diamond}\right)^{\rho_\lambda}$$

$$\rho_\lambda := \frac{1}{2}\left[\left(1 - \frac{2\left(g_0 - \Gamma^\dagger\right)}{\sigma_0^2}\right) + \sqrt{\left(1 - \frac{2\left(g_0 - \Gamma^\dagger\right)}{\sigma_0^2}\right)^2 + \frac{8}{\sigma_0^2}\left(\mu_0 + \lambda\right)}\right]$$

$$\begin{aligned}
\mathbf{f}_\lambda = & -\int_0^\infty da \int_0^{x^\diamond} dy \int_0^a ds\ \exp\{-\lambda(a-s)\} \int_0^{x^\diamond} dx \left[\frac{\partial}{\partial x} x P^\dagger(s, x)\right] \\
& K^\dagger(s, x \to a, y)\, F_s(y) \\
= & -\int_0^\infty da \int_0^{x^\diamond} dy \int_0^a ds\ \exp\{-\lambda(a-s)\} \int_0^{x^\diamond} dx \left[\frac{\partial}{\partial x} x P^\dagger(s, x)\right] \\
& \underbrace{K^\dagger(0, x \to a-s, y)}_{\text{the Strong Markov Property}} \quad F_s(y)
\end{aligned}$$

$$= -\int_0^{x^\circ} dx\ \psi_\lambda\left(x, \Gamma^\dagger\right) \frac{\partial}{\partial x}\left(x \hat{P}_0^\dagger(x)\right), \quad \hat{P}_0^\dagger(x) := \int_0^\infty da\ P^\dagger(a, x)$$

$$= \int_0^{x^\circ} dx\ \hat{P}_0^\dagger(x)\, x \frac{\partial}{\partial x} \psi_\lambda\left(x, \Gamma^\dagger\right) \Leftarrow \left[\psi_\lambda\left(x, \Gamma^\dagger\right) x \hat{P}_0^\dagger(x)\right]_{x=0}^{x=x^\circ} = 0,$$

$$\mathbf{T}_\lambda = -\gamma_0 \int_0^\infty da \int_0^{x^\circ} dy \int_0^a ds\ \exp\{-\lambda(a-s)\} \int_0^{x^\circ} dx \left[\frac{\partial}{\partial x} x P^\dagger(s, x)\right]$$
$$K^\dagger(0, x \to a-s, y)$$

$$= -\gamma_0 \int_0^{x^\circ} dy \int_0^{x^\circ} dx \frac{\partial}{\partial x}\left(x \hat{P}_0^\dagger(x)\right) \hat{K}^\dagger(0, x \to \lambda, y)$$

$$= -\frac{\gamma_0}{\mu_0 + \lambda} \int_0^{x^\circ} dx \frac{\partial}{\partial x}\left(x \hat{P}_0^\dagger(x)\right)\left(1 - \mathbb{E}_{X_0=x}^\dagger\left[\exp\left\{-(\mu_0+\lambda) a^\circ\right\}\right]\right)$$
$$\Leftarrow \text{see Eq. (86)}$$

$$= -\frac{\gamma_0}{\mu_0 + \lambda} \int_0^{x^\circ} dx \frac{\partial}{\partial x}\left(x \hat{P}_0^\dagger(x)\right)\left(1 - \left(\frac{x}{x^\circ}\right)^{\rho_\lambda}\right)$$

$$= -\frac{\gamma_0}{\mu_0 + \lambda} \int_0^{x^\circ} dx\ \hat{P}_0^\dagger(x)\, x \frac{\partial}{\partial x}\left(\frac{x}{x^\circ}\right)^{\rho_\lambda},$$

$$\mathbf{w}_\lambda^\dagger = \frac{\gamma_0}{\mu_0 + \lambda}\left(1 - \mathbb{E}_{X_0=x}^\dagger\left[\exp\left\{-(\mu_0+\lambda) a^\circ\right\}\right]\right)$$

$$= \frac{\gamma_0}{\mu_0 + \lambda}\left(1 - \left(\frac{x}{x^\circ}\right)^{\rho_\lambda}\right).$$

Putting the pieces together, the characteristic equation becomes

$$1 = \psi_\lambda\left(x, \Gamma^\dagger\right) - \frac{\frac{\gamma_0}{\mu_0+\lambda}\left(1 - \left(\frac{x}{x^\circ}\right)^{\rho_\lambda}\right) \int_0^{x^\circ} d\xi\ \hat{P}_0^\dagger(\xi)\, \xi \frac{\partial}{\partial \xi} \psi_\lambda\left(\xi, \Gamma^\dagger\right)}{1 - \frac{\gamma_0}{\mu_0+\lambda} \int_0^{x^\circ} d\xi\ \hat{P}_0^\dagger(\xi)\, \xi \frac{\partial}{\partial \xi}\left(\frac{\xi}{x^\circ}\right)^{\rho_\lambda}}$$

$$= \psi_\lambda\left(x, \Gamma^\dagger\right) - \frac{\frac{\gamma_0 \rho_\lambda}{\mu_0+\lambda}\left(1 - \left(\frac{x}{x^\circ}\right)^{\rho_\lambda}\right) \int_0^{x^\circ} d\xi\ \hat{P}_0^\dagger(\xi)\, \psi_\lambda\left(\xi, \Gamma^\dagger\right)}{1 - \frac{\gamma_0 \rho_\lambda}{\phi(x^\circ)(\mu_0+\lambda)} \int_0^{x^\circ} d\xi\ \hat{P}_0^\dagger(\xi)\, \psi_\lambda\left(\xi, \Gamma^\dagger\right)}.$$

Moreover, because of the Markov property, the following computation is valid.

$$\int_0^{x^\circ} d\xi\ \hat{P}_0^\dagger(\xi)\, \psi_\lambda\left(\xi, \Gamma^\dagger\right) = \frac{\Gamma^\dagger}{\gamma_0 e_0} \int_0^\infty da \int_0^a ds\, \exp\{-\lambda(a-s)\}$$
$$(e_0 : \text{see Eqs. (86) and (117).})$$
$$\times \int_0^{x^\circ} dy \int_0^{x^\circ} d\xi\ K^\dagger(0, x \to s, \xi)\, K^\dagger$$
$$(0, \xi \to a-s, y)\, F_S(y)$$

$$= \frac{\Gamma^\dagger}{\gamma_0 e_0} \int_0^\infty da \int_0^a ds \exp\{-\lambda (a-s)\}$$
$$\times \int_0^{x^\diamond} dy \int_0^{x^\diamond} d\xi\ K^\dagger (0, x \to s, \xi)\ K^\dagger$$
$$(s, \xi \to a, y)\ F_S(y)$$
$$= \frac{\Gamma^\dagger}{\gamma_0 e_0} \int_0^\infty da \int_0^a ds \exp\{-\lambda (a-s)\}$$
$$\times \int_0^{x^\diamond} dy\ K^\dagger (0, x \to a, y)\ F_S(y)$$
$$= \frac{\Gamma^\dagger}{\gamma_0 e_0} \int_0^\infty da \frac{1}{\lambda} (1 - \exp\{-\lambda a\})\ \mathbb{E}^\dagger_{X_0 = x}$$
$$\times \left[F_S(X_a) \exp\left\{ - \int_0^a \mu_S(X_\tau) \right\} \right]$$
$$= \frac{\Gamma^\dagger}{\gamma_0 e_0 \lambda} \left[1 - \psi_\lambda\left(x, \Gamma^\dagger\right)\right] \Leftarrow \psi_0\left(x, \Gamma^\dagger\right) = 1.$$

The characteristic equation is given by

$$1 = \psi_\lambda\left(x, \Gamma^\dagger\right) - \frac{\frac{\gamma_0 \rho_\lambda}{\mu_0 + \lambda}\left(1 - \left(\frac{x}{x^\diamond}\right)^{\rho_\lambda}\right) \frac{\Gamma^\dagger}{\gamma_0 e_0 \lambda}\left(1 - \psi_\lambda\left(x, \Gamma^\dagger\right)\right)}{1 - \frac{\gamma_0 \rho_\lambda}{\phi(x^\diamond)(\mu_0 + \lambda)} \frac{\Gamma^\dagger}{\gamma_0 e_0 \lambda}\left(1 - \psi_\lambda\left(x, \Gamma^\dagger\right)\right)}$$

$$\Leftrightarrow$$

$$1 - \psi_\lambda\left(x, \Gamma^\dagger\right) = -\frac{\frac{\gamma_0 \rho_\lambda}{\mu_0 + \lambda}\left(1 - \left(\frac{x}{x^\diamond}\right)^{\rho_\lambda}\right) \frac{\Gamma^\dagger}{\gamma_0 e_0 \lambda}\left(1 - \psi_\lambda\left(x, \Gamma^\dagger\right)\right)}{1 - \frac{\gamma_0 \rho_\lambda}{\phi(x^\diamond)(\mu_0 + \lambda)} \frac{\Gamma^\dagger}{\gamma_0 e_0 \lambda}\left(1 - \psi_\lambda\left(x, \Gamma^\dagger\right)\right)}$$

$$\Leftrightarrow$$

$$1 = -\frac{\frac{\gamma_0 \rho_\lambda}{\mu_0 + \lambda}\left(1 - \left(\frac{x}{x^\diamond}\right)^{\rho_\lambda}\right) \frac{\Gamma^\dagger}{\gamma_0 e_0 \lambda}}{1 - \frac{\gamma_0 \rho_\lambda}{\phi(x^\diamond)(\mu_0 + \lambda)} \frac{\Gamma^\dagger}{\gamma_0 e_0 \lambda}\left(1 - \psi_\lambda\left(x, \Gamma^\dagger\right)\right)}$$
$$= -\frac{\gamma_0 \rho_\lambda \Gamma^\dagger\left(\phi\left(x^\diamond\right) - \psi_\lambda\left(x, \Gamma^\dagger\right)\right)}{\phi\left(x^\diamond\right)\left(\mu_0 + \lambda\right) \gamma_0 e_0 \lambda - \gamma_0 \rho_\lambda \Gamma^\dagger\left(1 - \psi_\lambda\left(x, \Gamma^\dagger\right)\right)}$$
$$= -\frac{\mu_0 \rho_\lambda \Gamma^\dagger\left(\phi\left(x^\diamond\right) - \psi_\lambda\left(x, \Gamma^\dagger\right)\right)}{\left(\mu_0 + \lambda\right)\left(\phi\left(x^\diamond\right) - 1\right) \lambda - \mu_0 \rho_\lambda \Gamma^\dagger\left(1 - \psi_\lambda\left(x, \Gamma^\dagger\right)\right)}$$

$$\Leftrightarrow$$

$$\left(\mu_0 + \lambda\right)\left(\phi\left(x^\diamond\right) - 1\right)\lambda - \mu_0 \rho_\lambda \Gamma^\dagger\left(1 - \psi_\lambda\left(x, \Gamma^\dagger\right)\right) = -\mu_0 \rho_\lambda \Gamma^\dagger\left(\phi\left(x^\diamond\right) - \psi_\lambda\left(x, \Gamma^\dagger\right)\right)$$

$$\Leftrightarrow$$

$$\left(\mu_0 + \lambda\right)\lambda + \mu_0 \rho_\lambda \Gamma^\dagger = 0.$$

Because ρ_λ satisfies (see Eq. (2.55) and Eq. (3.34))

$$\rho_\lambda \left(\frac{\sigma_0^2}{2} \rho_\lambda + g_0 - \Gamma^\dagger - \frac{\sigma_0^2}{2} \right) - \mu_0 - \lambda = 0,$$

the characteristic equation is deformed as

$$\begin{aligned} &\lambda \left(\frac{\sigma_0^2}{2} \rho_\lambda + g_0 - \Gamma^\dagger - \frac{\sigma_0^2}{2} \right) + \mu_0 \Gamma^\dagger = 0 \\ \Leftrightarrow \\ &\lambda \left(\sqrt{\left(1 - \frac{2\left(g_0 - \Gamma^\dagger\right)}{\sigma_0^2}\right)^2 + \frac{8}{\sigma_0^2}\left(\mu_0 + \lambda\right)} - \left(1 - \frac{2\left(g_0 - \Gamma^\dagger\right)}{\sigma_0^2}\right) \right) + \frac{4\mu_0\Gamma^\dagger}{\sigma_0^2} = 0. \end{aligned} \tag{3.37}$$

Because this equation is monotonically increasing in λ, it does not have positive real roots. Moreover, by rewriting the equation as

$$\left(\sqrt{\left(1 - \frac{2\left(g_0 - \Gamma^\dagger\right)}{\sigma_0^2}\right)^2 + \frac{8}{\sigma_0^2}\left(\mu_0 + \lambda\right)} \right)^2 = \left[\left(1 - \frac{2\left(g_0 - \Gamma^\dagger\right)}{\sigma_0^2}\right) - \frac{4\mu_0\Gamma^\dagger}{\lambda\sigma_0^2} \right]^2$$

and rearranging the equation above for the orders of λ, we get

$$\lambda^3 + \mu_0\lambda^2 + \mu_0\Gamma^\dagger \left(1 - \frac{2\left(g_0 - \Gamma^\dagger\right)}{\sigma_0^2}\right)\lambda - 2\left(\frac{\mu_0\Gamma^\dagger}{\sigma_0}\right)^2 = 0. \tag{3.38}$$

Although Eq. (3.38) has a positive real root, it does not share the roots of Eq. (3.37). If Eq. (3.38) does not have complex roots because it has a non-negative real part, the equilibrium is stable. To prove this, we use *reductio ad absurdum*. Suppose, by contrast, that $c \pm id$ with $c > 0$ are roots of Eq. (3.38). The equation can then be written as

$$\begin{aligned} &(\lambda - b)(\lambda - c - id)(\lambda - c + id) = 0 \\ \Leftrightarrow \\ &\lambda^3 - (b + 2c)\lambda^2 + \left(2bc + c^2 + d^2\right)\lambda - b\left(c^2 + d^2\right) = 0. \end{aligned}$$

Here, $b + 2c$ is positive; hence, this equation does not correspond to Eq. (3.38). Therefore, Eq. (3.37) only has roots with negative real parts.

By proving the local stability, the stationary population has a meaning that is identical to the *carrying capacity*. Thus, the *carrying capacity* is constrained by vital rates and intra-specific competition as internal factors (not external factors). When we consider the K-selection theory, such a simple model will be suggestive.

References

MacArthur, R., & Wilson, E. (1967). *The theory of island biogeography*. Princeton, NJ: Princeton Univ Press.

Pianka, E. (1976). Natural selection of optimal reproductive tactics. *American Zoologist, 16*(4), 775–784.

Reznick, D., Bryant, M. J., & Bashey, F. (2002). r-and k-selection revisited: the role of population regulation in life-history evolution. *Ecology, 83*(6), 1509–1520.

Allee, W., & Bowen, E. S. (1932). Studies in animal aggregations: mass protection against colloidal silver among goldfishes. *Journal of Experimental Zoology, 61*(2), 185–207.

Oizumi, R., Kuniya, T., & Enatsu, Y. (2016). Reconsideration of r/k selection theory using stochastic control theory and nonlinear structured population models. *PloS One, 11*(6), e0157715.

Chapter 4
Life History Evolution and Adaptive Stochastic Controls

Thus far, we have examined the properties of population dynamics generated by the life history through linear and non-linear age–multi-state structured models. In some cases, the life history parameters determine the carrying capacity and intrinsic rate of natural increase. Furthermore, it is thought that these life history parameters of living organisms are optimized in the process of evolution.

When such optimization is discussed as a mathematical science, it involves the maximization or minimization of some objective function. This objective function is called fitness, and evolution is said to work in the direction of maximizing this fitness. In classical evolution theory, fitness refers to the rate of population growth, such as the intrinsic rate of natural increase. However, in the K-selection theory and genetics, which have a density effect, the degree of reproductive success, which refers to the BRN, is sometimes called fitness.

In this chapter, based on the discussions in the previous chapters, we explore the concept of fitness that can be widely applied to connect life history and population dynamics.

4.1 Basic Theorem for Adaptive Life History

Two pioneering studies had attempted to discuss evolution from the perspective of the relationship between life history and population dynamics (Taylor et al., 1974; Leon, 1976). The basis of these studies is the r-selection theory, which considers the intrinsic rate of natural increase as fitness. In general, the life history that maximizes the number of births per individual is insufficient to maximize the population growth rate. Shortening the generation time by precocity is also a factor in increasing the population growth. The following function can simultaneously satisfy these requirements: the BRN with the discount rate for each age.

R. Oizumi, *Population Dynamics Based on Individual Stochasticity*,
Population Studies of Japan,
https://doi.org/10.1007/978-981-19-3548-0_4

$$\varphi(r) = \int_0^{\alpha} da \exp\{-ra\} \underbrace{F(a)}_{\text{fertility}} \times \underbrace{l(a)}_{\text{survivorship}} .$$

This function constitutes the characteristic equation of the McKendrick equation (Metz and Diekmann, 1986; Inaba, 2017).

Let the set of life history control be $\mathbb{U}$, e.g., allocate energy from foraged resources to reproduction or increase the survival rate. When $u(a) \in \mathbb{U}$ is an element of the control set at age a, then, if there exists a control $\tilde{u}_r$ that satisfies the following equation, it is the adaptive control of life history called the r-strategy.

$$\varphi[\tilde{u}_r](r) = \sup_{u \in \mathbb{U}} \left\{ \int_0^{\alpha} da \exp\{-ra\} F(a, u(a)) l(a, u(a)) \right\}, \tag{4.1}$$

where $f[u](\cdot)$ represents a functional for $u(a)$. Meanwhile, the objective function in K-selection was defined as follows by using the *carrying capacity* K as an external parameter (Leon, 1976):

$$\varphi[\tilde{u}_K](K) = \sup_{u \in \mathbb{U}} \left\{ \int_0^{\alpha} da\, F(a, u(a), K) l(a, u(a), K) \right\}. \tag{4.2}$$

The saturated population has $r = 0$. By assuming that $\varphi[\tilde{u}_K](K)$ was monotonically decreasing in K, the study modeled the adaptive control $\tilde{u}_K$ maximizing the *carrying capacity* K, which is called the K-strategy, similarly to the conventional r/K selection theory.

There are two perspectives on the discussion of the r/K selection theory. One is associated with the categorization of $\tilde{u}_r$ and $\tilde{u}_K$. According to (Pianka, 1976), $\tilde{u}_r$ tends to have precocity, prolificacy, and high mortality, whereas, $\tilde{u}_K$ has altriciality, low birth, and low mortality. However, by observing the life histories of real organisms, such a simple categorization is difficult and the convincing classification of $\tilde{u}_K$ is still not sufficient (Reznick et al., 2002). From the mathematical point of view, $\varphi[\tilde{u}_K](K)$ does not have a basis to support Pianka's classification except for maximizing the carrying capacity.

The other perspective concerns the maximization of the carrying capacity for adaptation. The carrying capacity is defined as the maximum population size in the environment where an organism lives. When the population size reaches the carrying capacity, the BRN is equal to one. However, the maximum population size does not always control the reproduction number. It is determined by the strength of the density effect, i.e., it is not K but Γ. We have already treated this in the previous chapter. Compared with Eqs. (3.16), eqrefks is a special case of $\gamma_\ell = 1$.

Now, we reconstruct the previous basic theorem based on the multi-state structured model. The objective function corresponding with Eq. (4.1) is Eq. (2.24) in age–multi-state structured population models. Furthermore, in K–selection, the objective function given by Eq. (4.2) is incorporated in Eq. (3.28). Let us consider the following function that unifies the objective functions in r-selection and K-selection.

$$\psi_r[u](\Gamma) := \int_0^\alpha da \exp\{-ra\} \int_A dy \int_A dx\, v(x) K(0, x \to a, y; \Gamma) F(a, y, u(a), \Gamma)$$
$$= \int_0^\alpha da \int_A dx\, v(x) \mathbb{E}_{X_0=x}\left[F(a, X_a, u(a), \Gamma) \exp\left\{-\int_0^a d\tau\, (\mu(\tau, X_\tau, u(\tau), \Gamma) + r)\right\}\right]. \tag{4.3}$$

In r-selection, the intrinsic rate of natural increase r_0 for the population whose life histories are controlled by $u(a)$ is derived from

$$\psi_r[u](\Gamma)\Big|_{r=r_0, \Gamma=\mathbf{o}} = 1,$$

where $\mathbf{o}$ is an m-dimensional 0 vector. Alternatively, the strength of the density effects generating the stationary population instead of r_0 satisfies

$$\psi_r[u](\Gamma)\Big|_{r=0, \Gamma=\Gamma^\dagger} = 1.$$

If $\tilde{u}(a, X_a) \in \mathbb{U}$ is the adaptive control of the life schedules, it should satisfy the following theorem.

Theorem 1 *Let* $\tilde{u} = \tilde{u}(a, X_a, r, \Gamma)$ *be*

$$\tilde{u} \in \mathbb{U}, \quad s.t. \quad r_0\left[\tilde{u}\right](\Gamma) = \sup_{u \in \mathbb{U}} r_0[u](\Gamma),$$

Let $\psi_r[u](\Gamma)$ *be given by Eq.* (4.3) *and* $\psi_{r_0[u]}[u](\Gamma) = 1$. *We define* $\tilde{r}_0(\Gamma)$ *as* $\tilde{r}_0 := r_0\left[\tilde{u}\right](\Gamma)$ *as follows: Then, we have*

$$\psi_{\tilde{r}_0}[u](\Gamma) \le \psi_{\tilde{r}_0}\left[\tilde{u}\right](\Gamma) = 1.$$

This theorem is easily verified because of the monotonicity of $\psi_r[u](\Gamma)$ with respect to r. The theorem implies that a control that maximizes $\psi_r[u](\Gamma)$ is equivalent to maximizing the dominant characteristic root $r_0(\Gamma)$ as a function of Γ (cf. (Oizumi et al., 2016)).

This theorem leads to two types of arguments: Let the maximized $\psi_r[u](\Gamma)$ be given by

$$\tilde{\psi}_r(\Gamma) := \sup_{u \in \mathbb{U}} \{\psi_r[u](\Gamma)\}. \tag{4.4}$$

One argument is related to the r-selection theory that maximizes the dominant characteristic root when we choose the following condition:

$$\tilde{\psi}_r(\Gamma)\Big|_{r=\tilde{r}_0, \Gamma=\mathbf{o}} = 1. \tag{4.5}$$

Because Γ represents the strength of the density effects, $\Gamma = \mathbf{o}$ denotes the adaptive strategy that will satisfy the selection of r. The second argument represents the conditions in K selection:

$$\psi_r [u] (\Gamma)\Big|_{r=0,\Gamma=\tilde{\Gamma}} \le \tilde{\psi}_r (\Gamma)\Big|_{r=0,\Gamma=\tilde{\Gamma}} = 1,$$

because the adaptive strategy in a stationary population is believed to be uninvaded by any strategy. Further, $\psi_0 [u] (\Gamma)$ is essentially the BRN, and therefore,

$$\tilde{\psi}_r (\Gamma)\Big|_{r=0,\Gamma=\tilde{\Gamma}} = 1 \tag{4.6}$$

is the necessary and sufficient condition for the adaptive strategy in K-selection (K-strategy) if and only if $\tilde{\Gamma}$ has negative real parts in all characteristic roots of Eq. (3.29).

Thus, the theorem argues that both the r-strategy and the K-strategy maximize identical objective function. In particular, the K-strategy maximizes the intrinsic rate of natural increase as a zero value, resulting in maximized density effects. In other words, the selection pressure on traits of the life history does not vary in response to the strength of the density effects; the K-strategy maximizes the carrying capacity as a by-product of the maximization process of the population growth rate under K-selection. Note that maximizing the density effect does not always maximize the carrying capacity. Hereafter, we refer to Theorem 1 as the basic theorem of adaptive life schedule.

4.2 Optimal Stopping Problem in Life Histories

We have obtained the appropriate objective function as the fitness in the previous section. From a practical perspective, we provide an application associated with adaptive maturity in this section.

For semelparous species, optimal maturity timing is essential in their life schedule because they have only one opportunity to reproduce in their entire life. Spending a long time to become mature involves a high risk of accidental death, whereas precocity reduces the possibility of storing sufficient resources to reproduce. Unlike deterministic state growth, individual stochasticity has various mature ages to reach an identical state. When the mature age $a^{\diamond}$ is the first reaching age to a mature state, we consider the adaptive mature age $\tilde{a}$ that gives

$$\begin{aligned}\tilde{\varphi}_r(x,\Gamma) &:= \mathbb{E}_{X_0=x}\left[\phi(X_{\tilde{a}},\Gamma)\exp\left\{-\int_0^{\tilde{a}} da\ (\mu(X_a,\Gamma)+r)\right\}\right]\\ &= \sup_{a^\diamond}\left\{\mathbb{E}_{X_0=x}\left[\phi(X_{a^\diamond},\Gamma)\exp\left\{-\int_0^{a^\diamond} da\ (\mu(X_a,\Gamma)+r)\right\}\right]\right\}.\end{aligned} \tag{4.7}$$

From the discussion in the previous section of this chapter, Eq. (4.7) constitutes the adaptive objective function:

$$\tilde{\psi}_r(\Gamma) = \int_D dx\ \nu(x)\,\tilde{\varphi}_r(x,\Gamma),$$

where

$$D := \left\{x \in A \middle| \phi(x,\Gamma) < \tilde{\varphi}_r(x,\Gamma)\right\}.$$

When each state satisfies D, individuals should continue to grow, which is called the *continuation region* (Øksendal, 2003). Because it is unrealistic for the initial state to reach the mature state, they should take values of the continuation region in nature.

For simplicity, we deal with an age-homogeneous system such that

$$dX_a^j = g_j(X_a,\Gamma)\,da + \sum_{k=1}^{n} \sigma_{jk}(X_a,\Gamma)\,dB_a^k,\quad X_s^j = x^j,\quad \sigma_{jk}(X_a,\Gamma) \geq 0$$

$$j = 1,2,\ldots,d. \tag{4.8}$$

$$\mu(x,\Gamma) > 0, \tag{4.9}$$

$$F_s(X_a,\Gamma) := \begin{cases} 0 & a < a^\diamond \\ \phi(X_a,\Gamma) > 0 & a = a^\diamond \end{cases}. \tag{4.10}$$

Let $\varphi_r(x,\Gamma)$ be

$$\varphi_r(x,\Gamma) := \mathbb{E}_{X_0=x}\left[\phi(X_{a^\diamond},\Gamma)\exp\left\{-\int_0^{a^\diamond} da\ (\mu(X_a,\Gamma)+r)\right\}\right].$$

If the mature age is statistically bounded, $\mathbb{E}_{X_0=x}\left[a^\diamond\right] < \infty$, and $\phi(x,\Gamma)$ is a second-order differentiable function for x, the following representation is known as Dynkin's formula (Karatzas and Shreve, 1991; Øksendal, 2003):

$$\begin{aligned}\varphi_r(x,\Gamma) = \phi(x,\Gamma) + \mathbb{E}_{X_0=x}\Bigg[\int_0^{a^\diamond} da\ &\left(r - H^*(X_a,\Gamma)\right)\phi(X_a,\Gamma)\\ &\times\exp\left\{-\int_0^{a^\diamond} d\tau\ (\mu(X_\tau,\Gamma)+r)\right\}\Bigg],\end{aligned} \tag{4.11}$$

where

$$H^{*}(x,\Gamma)\phi(x,\Gamma)=\mu(x,\Gamma)\phi(x,\Gamma)-\sum_{j=1}^{d}g_{j}(x,\Gamma)\frac{\partial}{\partial x^{j}}\phi(x,\Gamma)$$
$$-\frac{1}{2}\sum_{j=1}^{d}\sum_{j'=1}^{d}S_{jj'}(x,\Gamma)\frac{\partial^{2}}{\partial x^{j}\partial x^{j'}}\phi(x,\Gamma). \tag{4.12}$$

Focusing on the second term on the right-hand side of Eq. (4.11), if $\phi(x,\Gamma)$ is

$$\left(r-H^{*}(x,\Gamma)\right)\phi(x,\Gamma)>0, \tag{4.13}$$

each individual should not reproduce at the state x. Recalling the Feynman–Kac formula, it derives the following equation.

$$\lim_{a\downarrow 0}\frac{1}{a}\left(\mathbb{E}_{X_{0}=x}\left[\phi(X_{a},\Gamma)\exp\left\{-\int_{0}^{a}d\tau\ (\mu(X_{\tau},\Gamma)+r)\right\}\right]-\phi(x,\Gamma)\right)$$
$$=\left(r-H^{*}(x,\Gamma)\right)\phi(x,\Gamma).$$

Because Eq. (4.13) implies that the state growth increases the expectation of the objective function, individuals at state x should continue to grow. Therefore, if Eq. (4.13) is valid in all x, the adaptive mature age does not exist. For adaptive maturity to exist, $\phi(x,\Gamma)$ should contain a region that gives

$$\left(r-H^{*}(x,\Gamma)\right)\phi(x,\Gamma)\leq 0. \tag{4.14}$$

When a state x satisfies this inequality, the individuals should switch to reproducing immediately. Adaptive semelparous species spend their juvenile states in D and must reproduce at the moment the state $x^{\diamond}$ satisfies Eq. (4.14). In addition, the mature state also satisfies the following inequality for all the mature ages.

$$\phi\left(x^{\diamond},\Gamma\right)\geq\mathbb{E}_{X_{0}=x^{\diamond}}\left[\phi(X_{a^{\diamond}},\Gamma)\exp\left\{-\int_{0}^{a^{\diamond}}d\tau\ (\mu(X_{\tau},\Gamma)+r)\right\}\right],\ \forall a^{\diamond}. \tag{4.15}$$

Thus, the individuals already reach the adaptive mature state. If $\phi(x,\Gamma)$ holds for Eqs. (4.14) and (4.15) in $x^{\diamond}$, it is called *the superharmonic (mean-valued) majorant* and it coincides with $\tilde{\varphi}_{r}(x,\Gamma)$ (Øksendal, 2003) such that

$$\phi\left(x^{\diamond},\Gamma\right)=\tilde{\varphi}_{r}\left(x^{\diamond},\Gamma\right).$$

4.3 Adaptive Reproductive Timing for Semelparous Species

Let us consider determining the adaptive maturity through a concrete example. Here, we reuse the *resource acquisition competition model* (see 3.4.1). The following objective function constitutes the characteristic equation of this model:

$$\psi_r(x,\Gamma) = \phi\left(x^{\diamond}\right)\left(\frac{x}{x^{\diamond}}\right)^{\rho_r(\Gamma)} \tag{4.16}$$

$$\rho_r(\Gamma) := \frac{1}{2}\left[\left(1-\frac{2(g_0-\Gamma)}{\sigma_0^2}\right)+\sqrt{\left(1-\frac{2(g_0-\Gamma)}{\sigma_0^2}\right)^2+\frac{8}{\sigma_0^2}(\mu_0+r)}\right]. \tag{4.17}$$

$\psi_r(y,0)$ is associated with Eq. (2.57), and $\psi_0\left(y,\Gamma^{\dagger}\right)$ is equal to the solution of Eq. (3.32). Let us take the logarithm of both sides of Eq. (4.16):

$$\ln\{\psi_r(x,\Gamma)\} = \ln\left\{\phi\left(x^{\diamond}\right)\right\} - \rho_r(\Gamma)\ln\left\{\frac{x^{\diamond}}{x}\right\}.$$

In the absence or presence of the density effect, the characteristic equation becomes

$$\ln\{\psi_r(x,\Gamma)\}\Big|_{r=r_0,\Gamma=0} = \ln\{\psi_r(x,\Gamma)\}\Big|_{r=0,\Gamma=\Gamma^{\dagger}} = 0.$$

In other words, the following relation is valid.

$$\frac{\ln\left\{\phi\left(x^{\diamond}\right)\right\}}{\ln\left\{\frac{x^{\diamond}}{x}\right\}} = \rho_r(\Gamma)\Big|_{r=r_0,\Gamma=0} = \rho_r(\Gamma)\Big|_{r=0,\Gamma=\Gamma^{\dagger}}.$$

Because $\rho_r(\Gamma)$ is monotonically increasing in r and Γ, the adaptive mature size $\tilde{x}$ gives

$$\frac{\ln\{\phi(\tilde{x})\}}{\ln\left\{\frac{\tilde{x}}{x}\right\}} = \sup_{x^{\diamond}}\frac{\ln\left\{\phi\left(x^{\diamond}\right)\right\}}{\ln\left\{\frac{x^{\diamond}}{x}\right\}}. \tag{4.18}$$

If $\tilde{x} < \infty$, the adaptive mature age $\tilde{a}$ is the first reaching age to $\tilde{x}$.

This result implies that the adaptive mature age is determined only by the ratio of the birth rate to the mature size, with or without the density effect. In general, a large mature size has the advantage of retaining many germ cells, whereas it takes a long time to grow and involves the risk of accidental death. Thus, the trade-off between fertility and mortality determines the adaptive mature size. In this model, the adaptive mature age can maximize both the intrinsic rate of natural increases and the carrying capacity.

4.4 Optimal Life Schedule Problem and Hamilton–Jacobi–Bellman Equation

The growth, reproduction, and mortality that characterize the species are thought to be optimized during evolution. Analysis and modeling of the life schedule using control theory have a long history because the evolution of life histories is interpreted as the optimal control of the life schedule. For example, mathematical biologists have discussed the optimal switching timing from the growth of vegetative organs to reproductive organs in plants (Iwasa and Roughgarden, 1984; Iwasa and Cohen, 1989). Such an optimization problem is collectively referred to as the *optimal life schedule problem.* This section reconstructs the general theory for the optimal life schedule problem according to the basic theorem for the optimal life schedule.

Let us consider the ingredient model for life history with control parameters $u \in \mathbb{U}$.

$$dX_a^j = g_j\left(a, X_a, u(a), \Gamma\right) da + \sum_{k=1}^{n} \sigma_{jk}\left(a, X_a, u(a), \Gamma\right) dB_a^k, \quad X_0^j = x^j,$$
$$\sigma_{jk}\left(a, X_a, u(a), \Gamma\right) \geq 0 \quad j = 1, 2, \cdots, d, \tag{4.19}$$

$$\mu\left(a, y, u, \Gamma\right) > 0, \quad \int_0^{\alpha} da\, \mu\left(a, y, u, \Gamma\right) = +\infty \quad \text{for all } x \in A, \tag{4.20}$$

and

$$F\left(a, y, u, \Gamma\right) \geq 0. \tag{4.21}$$

Both r-selection and K-selection have constant density effects ($\Gamma = 0$ or $\Gamma^{\dagger} > 0$); hence, Γ is a time-independent non-negative vector value $\Gamma \in \mathbb{R}_+^m$.

In accordance with Sect. 4.1, we derive the adaptive control satisfying Theorem 1. Recalling the relationship between the reproductive value given by Eq. (2.35) and the characteristic function given by Eq. (2.36), the objective function given by Eq. (4.3) is supposed to consist of the following integrand (Oizumi and Inaba, 2021).

$$\psi_r\left[u\right]\left(\Gamma\right) = \int_A dx\, v_r[u]\left(0, x, \Gamma\right) \nu\left(x\right). \tag{4.22}$$

Then, $v_r[u]\left(0, x, \Gamma\right)$ is

$$v_r[u]\left(0, x, \Gamma\right) := \mathbb{E}_{X_a = y}\left[\int_a^{\alpha} d\tau\, F\left(\tau, X_\tau, u\left(\tau\right), \Gamma\right)\right.$$
$$\left. \times \exp\left\{-\int_a^{\tau} d\tau_1\, \left(\mu\left(\tau_1, X_{\tau_1}, u\left(\tau_1\right), \Gamma\right) + r\right)\right\}\right], \tag{4.23}$$

where the expectation $\mathbb{E}_{X_a = y}\left[\cdot\right]$ is given by the probability measure of Eq. (4.19). Let the maximized Eq. (4.23) be

$$\tilde{v}_r(a, y, \Gamma) := \sup_{u \in \mathbb{U}} \{v_r[u](0, x, \Gamma)\}. \tag{4.24}$$

It can constitute the maximized objective function Eq. (4.4) such that

$$\tilde{\psi}_r(\Gamma) = \int_A dx\, \tilde{v}_r(0, x, \Gamma)\, \nu(x). \tag{4.25}$$

To find the adaptive control, first, we obtain an equation that $\tilde{v}$ satisfies. Then, according to the control theory, $\tilde{v}$, which is called the value function, obeys *Bellman's principle* (or the *principle of optimality*) (Bellman, 2013):

"an optimal strategy has the property that whatever the initial state and initial control are, the remaining control must constitute an optimal strategy with regard to the state resulting from the first strategy."

Specifically, it implies the following relationship:

$$\begin{aligned}
&\tilde{v}_r(a_0, y, \Gamma) = \\
&\sup_{u \in \mathbb{U}} \left\{ \mathbb{E}_{X_a = y} \left[\tilde{v}_r(a, X_a, \Gamma) \exp\left\{ - \int_{a_0}^{a} d\tau\, \left(\mu(\tau, X_\tau, u(\tau), \Gamma) + r\right) \right\} \right. \right. \\
&\left.\left. + \int_{a_0}^{a} d\tau\, F(\tau, X_\tau, u(\tau), \Gamma) \exp\left\{ - \int_{a_0}^{\tau} d\tau_1\, \left(\mu(\tau_1, X_{\tau_1}, u(\tau_1), \Gamma) + r\right) \right\} \right] \right\},
\end{aligned} \tag{4.26}$$

where $0 \le a_0 \le a \le \alpha$. Eq. (4.26) implies that the adaptive control with the value function from age a to α as the terminal condition must optimize the objective function before age a. Eq. (4.26) can be deformed as

$$\begin{aligned}
0 = &\sup_{u \in \mathbb{U}} \left\{ \mathbb{E}_{X_a = y} \left[\tilde{v}_r(a, X_a, \Gamma) \exp\left\{ - \int_{a_0}^{a} d\tau\, \left(\mu(\tau, X_\tau, u(\tau), \Gamma) + r\right) \right\} - \tilde{v}_r(a_0, y, \Gamma) \right. \right. \\
&\left.\left. + \int_{a_0}^{a} d\tau F(\tau, X_\tau, u(\tau), \Gamma) \exp\left\{ - \int_{a_0}^{\tau} d\tau_1\, \left(\mu(\tau_1, X_{\tau_1}, u(\tau), \Gamma) + r\right) \right\} \right] \right\} \\
= &\sup_{u \in \mathbb{U}} \left\{ \mathbb{E}_{X_a = y} \left[\int_{a_0}^{a} d\left(\tilde{v}_r^*(\tau, X_\tau, \Gamma) \exp\left\{ - \int_{a_0}^{\tau} d\tau_1\, \left(\mu(\tau_1, X_{\tau_1}, u(\tau_1), \Gamma) + r\right) \right\} \right) \right. \right. \\
&\left.\left. + \int_{a_0}^{a} d\tau F(\tau, X_\tau, u(\tau), \Gamma) \exp\left\{ - \int_{a_0}^{\tau} d\tau'\, \left(\mu(\tau_1, X_{\tau_1}, u(\tau_1), \Gamma) + r\right) \right\} \right] \right\} \\
&(a \to a_0 + \delta a) \\
= &\sup_{u \in \mathbb{U}} \left\{ \mathbb{E}_{X_a = y} \left[\int_{a_0}^{a_0 + \delta a} d\tau\, \underbrace{\left[\frac{\partial}{\partial \tau} - H^*(\tau, X_\tau, u(\tau), \Gamma) - r \right] \tilde{v}_r(\tau, X_\tau, \Gamma)}_{\text{Feynman-Kac formula}} \right. \right. \\
&\times \exp\left\{ - \int_{a_0}^{\tau} d\tau_1\, \left(\mu(\tau_1, X_{\tau_1}, u(\tau), \Gamma) + r\right) \right\}
\end{aligned}$$

$$+\int_{a_0}^{a_0+\delta a} d\tau F(\tau, X_\tau, u(\tau), \Gamma) \exp\left\{-\int_{a_0}^{\tau} d\tau_1 \left(\mu\left(\tau_1, X_{\tau_1}, u(\tau_1), \Gamma\right)+r\right)\right\}\Bigg]\Bigg\}.$$

From the deformation above, we consider the limit

$$0 = \lim_{\delta a\downarrow 0}\frac{1}{\delta a}\sup_{u\in\mathbb{U}}\Bigg\{\mathbb{E}_{X_a=y}\Bigg[\int_{a_0}^{a_0+\delta a} d\tau\left[\frac{\partial}{\partial\tau}-H^*(\tau, X_\tau, u(\tau), \Gamma)-r\right]\tilde{v}_r(\tau, X_\tau, \Gamma)$$

$$\times\exp\left\{-\int_{a_0}^{\tau} d\tau_1 \left(\mu\left(\tau_1, X_{\tau_1}, u(\tau_1), \Gamma\right)+r\right)\right\}$$

$$+\int_{a_0}^{a_0+\delta a} d\tau F(\tau, X_\tau, u(\tau), \Gamma)\exp\left\{-\int_{a_0}^{\tau} d\tau_1 \left(\mu\left(\tau_1, X_{\tau_1}, u(\tau_1), \Gamma\right)+r\right)\right\}\Bigg]\Bigg\}$$

$$=\frac{\partial}{\partial a_0}\tilde{v}_r(a_0, y, \Gamma)-\inf_{u\in\mathbb{U}}\left\{\left[H^*(a_0, y, u, \Gamma)-r\right]\tilde{v}_r(a_0, y, \Gamma)-F(a_0, y, u, \Gamma)\right\}.$$

Shifting the age to $a_0 = a$, we have

$$\frac{\partial}{\partial a}\tilde{v}_r(a, y, \Gamma)-\inf_{u\in\mathbb{U}}\left\{\left[H^*(a, y, u, \Gamma)-r\right]\tilde{v}_r(a, y, \Gamma)-F(a, y, u, \Gamma)\right\}=0 \tag{4.27}$$

Equation (4.27) is referred to as the *Hamilton–Jacobi–Bellman equation (HJB equation)* in control theory (Yong and Zhou, 1999). Finding the adaptive control $\tilde{u}$ requires the existence of a value function satisfying Eqs. (4.27) and (4.5) or (4.6). Suppose that the arbitrary constant of the reproductive value is one.[1] Hence, we can find a similarity between the adjoint and HJB equations (Eqs. (2.33) and (4.27)). Therefore, the latter equation is interpreted as the adaptive control minimizing the decreasing rate of the value function constituting reproductive value. Moreover, because the reproductive value at age zero constitutes the characteristic function in this case, the adaptive control is identical to the adaptive strategy maximizing the objective function (Eq. (4.3)). Putting the pieces together, we get

$$\begin{aligned}
&\frac{\partial}{\partial a}\tilde{v}_r(a, y, \Gamma)=\inf_{u\in\mathbb{U}}\left\{\left[H^*(a, y, v, \Gamma)-r\right]\tilde{v}_r(a, y, \Gamma)-F(a, y, u, \Gamma)\right\}\\
&\quad\tilde{v}_r(\alpha, y, \Gamma)=0 \\
&\tilde{\psi}_r(\Gamma)\Big|_{r=\tilde{r}_0,\Gamma=\mathbf{0}}=1,\quad\text{or}\quad\tilde{\psi}_r(\Gamma)\Big|_{r=0,\Gamma=\tilde{\Gamma}}=1
\end{aligned}\tag{4.28}$$

The HJB equation, i.e., Eq. (4.28), implies that the function providing the reproductive value is the essential objective function as the fitness.

[1] $\int_A d\xi\ v(0,\xi)\,v(\xi)=1$ in Eq. (2.34).

4.4.1 Stationary Control

At the end of this section, we introduce a specific control in age-homogeneous life history. Let us consider the age-homogeneous state transition system Eq. (4.8) with the control parameter such that

$$\begin{aligned} dX_a^j &= g_j\left(X_a, u, \Gamma\right) da + \sum_{k=1}^{n} \sigma_{jk}\left(X_a, u, \Gamma\right) dB_a^k, \quad X_0^j = y^j, \\ \sigma_{jk}&\left(X_a, u, \Gamma\right) \geq 0 \quad j = 1, 2, \cdots, d. \end{aligned} \tag{4.29}$$

Suppose that individuals following Eq. (4.29) die accidentally only, i.e., $\alpha = +\infty$. Then, the mortality is also assumed to be homogeneous in age as follows:

$$\mu\left(y, u, \Gamma\right) > 0, \quad \forall u \in \mathbb{U}. \tag{4.30}$$

The fertility is also assumed to be independent of age a.

$$F\left(y, u, \Gamma\right) \geq 0. \tag{4.31}$$

Substituting Eqs. (4.29), (4.30), and (4.31) into the value function Eq. (4.24), we have

$$\begin{aligned} \tilde{v}_r\left(a, y, \Gamma\right) := \sup_{u \in \mathbb{U}} \Bigg\{ \mathbb{E}_{X_a = y} \Bigg[\int_a^{\alpha} d\tau \; F\left(X_\tau, u\left(\tau\right), \Gamma\right) \\ \exp\left\{ -\int_a^{\tau} d\tau_1 \; \left(\mu\left(X_{\tau_1}, u\left(\tau_1\right), \Gamma\right) + r\right) \right\} \Bigg] \Bigg\}, \end{aligned} \tag{4.32}$$

As the life history system (Eqs. (4.29), (4.30), and (4.31)) does not depend on age explicitly, it has the strong Markovian property. Hence, the age parameter in the equation above can be shifted as follows:

$$\begin{aligned} \tilde{v}_r\left(a, y, \Gamma\right) = \sup_{u \in \mathbb{U}} \Bigg\{ \mathbb{E}_{X_0 = y} \Bigg[\int_0^{\alpha - a} d\tau \; F\left(X_\tau, u\left(\tau\right), \Gamma\right) \\ \times \exp\left\{ -\int_0^{\tau} d\tau_1 \; \left(\mu\left(X_{\tau_1}, u\left(\tau_1\right), \Gamma\right) + r\right) \right\} \Bigg] \Bigg\} \end{aligned}$$

Taking the limit $\alpha \to \infty$, the value function does not depend on age:

$$\begin{aligned} \tilde{v}_r\left(a, y, \Gamma\right) = \lim_{\alpha \uparrow \infty} \sup_{u \in \mathbb{U}} \Bigg\{ \mathbb{E}_{X_0 = y} \Bigg[\int_0^{\alpha - a} d\tau \; F\left(X_\tau, u\left(\tau\right), \Gamma\right) \\ \times \exp\left\{ -\int_0^{\tau} d\tau_1 \; \left(\mu\left(X_{\tau_1}, u\left(\tau_1\right), \Gamma\right) + r\right) \right\} \Bigg] \Bigg\} \end{aligned}$$

$$
\begin{aligned}
&= \sup_{u \in \mathbb{U}} \left\{ \mathbb{E}_{X_0 = y} \left[\int_0^\infty d\tau \, F\left(X_\tau, u(\tau), \Gamma\right) \right.\right. \\
&\quad \left.\left. \times \exp\left\{ - \int_0^\tau d\tau_1 \, \left(\mu\left(X_{\tau_1}, u(\tau_1), \Gamma\right) + r \right) \right\} \right] \right\} \quad (4.33) \\
&= \tilde{v}_r(y, \Gamma) . \quad (4.34)
\end{aligned}
$$

From Bellman's principle, i.e., Eq. (4.26), the following relationship is valid in this case.

$$
\begin{aligned}
\tilde{v}_r(y, \Gamma) &= \sup_{u \in \mathbb{U}} \left\{ \mathbb{E}_{X_0 = y} \left[\int_0^a d\tau \, F\left(X_\tau, u(\tau), \Gamma\right) \exp\left\{ - \int_0^\tau d\tau_1 \, \left(\mu\left(X_{\tau_1}, u(\tau_1), \Gamma\right) + r \right) \right\} \right.\right. \\
&\quad \left.\left. + \int_a^\infty d\tau \, F\left(X_\tau, u(\tau), \Gamma\right) \exp\left\{ - \int_0^\tau d\tau_1 \, \left(\mu\left(X_{\tau_1}, u(\tau_1), \Gamma\right) + r \right) \right\} \right] \right\} \\
&= \sup_{u \in \mathbb{U}} \left\{ \mathbb{E}_{X_0 = y} \left[\int_0^a d\tau \, F\left(X_\tau, u(\tau), \Gamma\right) \exp\left\{ - \int_0^\tau d\tau_1 \, \left(\mu\left(X_{\tau_1}, u(\tau_1), \Gamma\right) + r \right) \right\} \right.\right. \\
&\quad \left.\left. + \tilde{v}_r(X_a, \Gamma) \exp\left\{ - \int_0^a d\tau_1 \, \left(\mu\left(X_{\tau_1}, u(\tau_1), \Gamma\right) + r \right) \right\} \right] \right\} .
\end{aligned}
$$

Similarly to the of derivation of Eq. (4.27),

$$
\begin{aligned}
0 &= \lim_{a \downarrow 0} \sup_{u \in \mathbb{U}} \left\{ \frac{1}{a} \mathbb{E}_{X_0 = y} \left[\int_0^a d\tau \, F\left(X_\tau, u(\tau), \Gamma\right) \exp\left\{ - \int_0^\tau d\tau_1 \, \left(\mu\left(X_{\tau_1}, u(\tau_1), \Gamma\right) + r \right) \right\} \right.\right. \\
&\quad \left.\left. - \int_0^a d\tau \, \left[H^*\left(X_\tau, u, \Gamma\right) + r \right] \tilde{v}_r(X_a, \Gamma) \exp\left\{ - \int_0^a d\tau_1 \, \left(\mu\left(X_{\tau_1}, u(\tau_1), \Gamma\right) + r \right) \right\} \right] \right\} ,
\end{aligned}
$$

and we have the HJB equation in the stationary control for age:

$$
0 = \inf_{u \in \mathbb{U}} \left\{ \left[H^*(y, u, \Gamma) + r \right] \tilde{v}_r(y, \Gamma) - F(y, u, \Gamma) \right\} , \quad (4.35)
$$

where

$$
\begin{aligned}
H^*(y, u, \Gamma)\, \phi(y) &= \mu(y, u, \Gamma)\, \phi(y) - \sum_{j=1}^d g_j(y, u, \Gamma) \frac{\partial}{\partial y^j} \phi(y) \\
&\quad - \frac{1}{2} \sum_{j=1}^d \sum_{j'=1}^d S_{jj'}(y, u, \Gamma) \frac{\partial^2}{\partial y^j \partial y^{j'}} \phi(y) .
\end{aligned}
$$

Such control, which is independent of age (or time), is called *stationary control.* Eq. (4.33) implies that the reproductive value depends only on the states and is independent of the age in this system, including Sects. 2.5 and 4.2.

4.5 Reconsideration of r/K Selection Theory

We analyzed how to find the adaptive life history strategy using a common objective function with or without the density effect. In contrast to previous studies on life history evolution, this book does not treat vital rates as direct variables of the population growth rate or reproductive success. By using the common characteristic function (Eq. (4.3)), these two indicators (r_0 and ψ_0) are maximized indirectly. This suggests that selection pressure affects identical objects. Early studies on evolution under r-selection used this characteristic function (Taylor et al., 1974). Because the characteristic function is always monotonically decreasing in r, the adaptive life histories tend to have precocity and prolificacy, which are the fundamental traits of r- selected species.

Meanwhile, K- selected species have the opposite traits, i.e., altriciality and low birth rate (Pianka, 1976). In nature, some species cannot be classified in these categories (Reznick et al., 2002). The conventional K-strategy maximizes the carrying capacity. In an early study using the characteristic equation for the K-strategy, it was assumed that the characteristic function was monotonically decreasing in the carrying capacity (Leon, 1976). Moreover, the carrying capacity was supposed to be an external parameter that is independent of the population structure, e.g., age, size, and other states. If we reconsider the carrying capacity as a comprehensive density effect, it is possible to evolve various life histories and not obsess over the r/K selection theory classification.

Our assertion does not contradict the conventional interpretation of the r/K-theory by the trade-off between survivorship and reproduction (Stearns, 1992). It provides more complicated scenarios for the K-strategy including the conventional one. As mentioned previously, the adaptive strategy under the density effect maximizes the density effect. It is not always related to the trade-off between surviving and reproducing. Here, we demonstrate such a case by using an analyzable model. Thus, the K-strategy can evolve more diverse life histories than we think. The density effect is a complicated factor in life history.

4.5.1 Adaptive Resource Utilization Model in Optimal Foraging Problem

Let us consider two resource acquisition competition models (see Sect. 3.4.1) whose size growth rates are statistically different from each other. One is given by

$$dX_a = (g_1 - \kappa\Gamma_t)\, X_a da + \sigma_1 X_a dB_a^1, \quad X_0 = x > 0. \tag{4.36}$$

The other is given by

$$dX_a = (g_2 - (1-\kappa)\,\Gamma_t)\, X_a da + \sigma_2 X_a dB_a^2, \quad X_0 = x > 0. \tag{4.37}$$

Each parameter has the following inequality:

$$g_1 > g_2 \geq 0, \quad \sigma_1 > \sigma_2 \geq 0.$$

If the density effect is sufficiently small, $\Gamma_t \approx 0$, Eq. (4.36) has a larger mean size growth rate and variance than Eq. (4.37). Further, $0 \leq \kappa \leq 1$ is a parameter that represents the symmetry of the density effects between the two models.

To avoid complications in the calculation, we use the following symbols:

$$\theta_\kappa^1(\Gamma) := (g_1 - \kappa\Gamma), \quad \theta_{1-\kappa}^2(\Gamma) := (g_2 - (1-\kappa)\Gamma). \tag{4.38}$$

Individuals consuming resource 1 follow Eq. (4.36) and those consuming resource 2 follow Eq. (4.37) in their size growth. When $u \in [0, 1]$ is the resource allocation rate, the size growth rate of both resource consumers obeys

$$dX_a = \left[\theta_\kappa^1(\Gamma)(1-u) - \theta_{1-\kappa}^2(\Gamma)u\right]X_a da + \left(\sigma_1(1-u)dB_a^1 + \sigma_2 u dB_a^2\right)X_a,$$
$$X_0 = x > 0. \tag{4.39}$$

As we only consider the stationary states for time t, Γ is set to be constant. Eq. (4.39) is an Ito-type Malthus equation (see Eq. (1.4)); hence, σ_1 decreases the sample paths reaching the mature size. In the absence of the density effect $\Gamma = 0$, this model is known as the *portfolio optimization problem* in financial engineering (e.g., (Aase, 1984; Øksendal, 2003; Shreve, 2004)). The problem is to find the optimal control for the appropriate risk aversion, which implies control of σ_1, in a portfolio of stocks or personal assets. Thus, the r-strategy in this model represents the optimal portfolio problem for foraging. A key difference from finance is that the optimal control maximizes the characteristic function Eq. (4.3) instead of the utility function.

The density effect also reduces the mean size growth rate in this model. As the stationary population is affected by the variance of the size growth representing individual stochasticity (see Eqs. (3.35) and (3.36)), the K-strategy should not ignore the control of the size growth statistics. Then, Eq. (3.35) indicates that σ_1 reduces the density effect. The effect of κ on the adaptive control is thought to be significant because it also affects the stationary population. We apply this size growth model to the semelparous model, which was analyzed in the sections of this and the previous chapters. Considering the adaptive growth curve in the semelparity will provide another perspective on the r/K-selection theory: let us consider the mortality and the fertility given by Eqs. (2.47) and (2.48). Equation (4.39) generates the following adjoint Hamiltonian:

$$H^*(y, u, \Gamma)\phi(y) =$$
$$\mu_0\phi(y) - \left[\theta_\kappa^1(\Gamma)(1-u) - \theta_{1-\kappa}^2(\Gamma)u\right]y\frac{\partial}{\partial y}\phi(y) - \frac{1}{2}\left(\sigma_1^2(1-u)^2 + \sigma_2^2u^2\right)y^2\frac{\partial^2}{\partial y^2}\phi(y)$$
$$y < x^\diamond. \tag{4.40}$$

This semelparous life history is an age-homogeneous system; hence, the HJB equation of this model is classified as stationary control such that

$$\inf_{u\in[0,1]}\left\{\left[H^{*}(x,u,\Gamma)+r\right]\tilde{v}_r(x,\Gamma)\right\}+F_{\mathrm{s}}(x)=0. \tag{4.41}$$

If the adaptive control $\tilde{u}$ exists in the range $0 < u < 1$, it must give an extremal value as a necessary condition:

$$\frac{\partial}{\partial u}\left(H^{*}(x,u,\Gamma)\tilde{v}_r(x,\Gamma)\right)\Big|_{u=\tilde{u}}=0.$$

By solving the equation above with respect to $\tilde{u}$, we get

$$\tilde{u}(\tilde{v})=\frac{\sigma_1^2}{\sigma_1^2+\sigma_2^2}+\frac{\left(\theta_\kappa^1(\Gamma)-\theta_{1-\kappa}^2(\Gamma)\right)\frac{\partial}{\partial x}\tilde{v}_r(x,\Gamma)}{\left(\sigma_1^2+\sigma_2^2\right)x\frac{\partial^2}{\partial x^2}\tilde{v}_r(x,\Gamma)}. \tag{4.42}$$

Substituting $\tilde{u}(\tilde{v})$ into

$$\left[H^{*}(x,u,\Gamma)+r\right]\tilde{v}_r(x,\Gamma)\Big|_{u=\tilde{u}(\tilde{v})}=0,\quad x<x^{\diamond}, \tag{4.43}$$

Equation (4.41) is rewritten as the following non-linear differential equation:

$$\begin{aligned}&-\left[\frac{\theta_\kappa^1(\Gamma)\sigma_2^2+\theta_{1-\kappa}^2(\Gamma)\sigma_1^2}{\sigma_1^2+\sigma_2^2}\right]x\frac{\partial}{\partial x}\tilde{v}_r(x,\Gamma)-\frac{1}{2}\frac{\sigma_1^2\sigma_2^2}{\sigma_1^2+\sigma_2^2}x^2\frac{\partial^2}{\partial x^2}\tilde{v}_r(x,\Gamma)\\&+\frac{1}{2}\frac{\left(\theta_\kappa^1(\Gamma)-\theta_{1-\kappa}^2(\Gamma)\right)^2\left(\frac{\partial}{\partial x}\tilde{v}_r(x,\Gamma)\right)^2}{\left(\sigma_1^2+\sigma_2^2\right)\frac{\partial^2}{\partial x^2}\tilde{v}_r(x,\Gamma)}+(\mu_0+r)\tilde{v}_r(x,\Gamma)=0,\quad x<x^{\diamond}.\end{aligned} \tag{4.44}$$

This equation seems to be a pathological form; however, it can be solved. Actually, the method to derive Eq. (2.57) is also useful in this equation. Putting an ansatz

$$\tilde{v}_r(x,\Gamma)=cx^{\tilde{\rho}},$$

into Eq. (4.44), we get a function similar to Eq. (2.57) as follows:

$$\begin{aligned}\tilde{v}_r(x,\Gamma)&=\phi\left(x^{\diamond}\right)\left(\frac{x}{x^{\diamond}}\right)^{\tilde{\rho}(r,\Gamma)}\\c&=\phi\left(x^{\diamond}\right)\left(x^{\diamond}\right)^{-\tilde{\rho}(r,\Gamma)},\quad 0<\tilde{\rho}(r,\Gamma).\end{aligned} \tag{4.45}$$

From Eqs. (2.58) and (3.34), $\tilde{\rho}(r,\Gamma)$ must be ρ_0 under both r and K selections.

$$\tilde{\rho}\left(\tilde{r}_0, 0\right) = \tilde{\rho}\left(0, \tilde{\Gamma}\right) = \rho_0 = \frac{\ln\left\{\phi\left(x^{\diamond}\right)\right\}}{\ln\left\{\frac{x^{\diamond}}{x}\right\}}. \tag{4.46}$$

Hence,[2] we can derive $\tilde{u}$ by substituting Eq. (4.45) into Eq. (4.42) without specifying $\tilde{\rho}$.

$$\tilde{u}\left(\Gamma\right) = \max\left\{\min\left\{\frac{\sigma_1^2}{\sigma_1^2 + \sigma_2^2} - \frac{g_1 - g_2 + (1 - 2\kappa)\,\Gamma}{\left(\sigma_1^2 + \sigma_2^2\right)(1 - \rho_0)}, 1\right\}, 0\right\}. \tag{4.47}$$

From a realistic point of view[3], we consider the case $\rho_0 < 1$ only. Equation (4.47) implies that the K-strategy coincides with the r-strategy in $\kappa = 0.5$. As for the control of the r-strategy, it becomes

$$\tilde{u}_r = \tilde{u}\left(0\right) = \max\left\{\frac{\sigma_1^2}{\sigma_1^2 + \sigma_2^2} - \frac{g_1 - g_2}{\left(\sigma_1^2 + \sigma_2^2\right)(1 - \rho_0)}, 0\right\}. \tag{4.48}$$

Then, the specialist using resource 2 only does not evolve because $\tilde{u}_r$ is always less than one (Oizumi, 2014). The K-strategy is more diverse than the other depending on κ and Γ because all strategies possibly evolve, i.e., resource 1 and 2 specialists or generalists using both resources (see Fig. 4.1). Thus, the K-strategy adjusts the evolution of the strategy to the density effect situation. Even this example suggests that the K-strategy is not always different from the r-strategy. The K-strategy must have a different strength of density effects from others in order to yield a difference from the r-strategy. $\kappa = 0.5$ does not generate the difference in the density effect.

4.5.2 Complicated Relationship among Demographic Indices under the Density Effect

The adaptive control in Eq. (4.39) does not influence the adaptive mature size. Hence, the growth rate and the probability of reaching the mature size are critical factors in the r/K strategies. The r-strategy optimizes these two to maximize r, and the K-strategy optimizes them in the stationary population. As an assertion of the conventional r/K selection theory, the r-strategy emphasizes reproduction, whereas the K-strategy emphasizes survivorship. If it is true, the K-strategy should provide a larger number of BRNs than the others, even if there is no density effect. To maximize r, maximizing the BRN is not necessary because the shortened generation time can augment the descendants per unit time.

---

[2] Equation (4.46) demonstrates that Eq. (4.18) is still valid and that the adaptive mature size is independent of the adaptive control in this model.

[3] Follow the interpretation of Eq. (2.59) in Sect. 2.5.

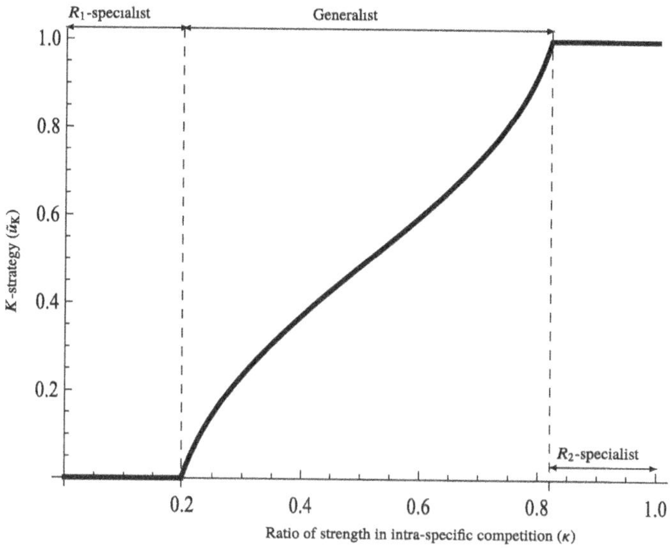

Fig. 4.1 Adaptive utilization under K-selection: this figure shows the shape of adaptive utilization under K-selection with respect to the ratio of the strength in the intra-specific competition κ. As the ratio increases, the adaptive strategy shifts to R_1-specialists, generalists, and R_2-specialists in turn. It is known that R_2-specialists do not evolve in the absence of density effects (Oizumi, 2014). The parameters are $g_1 = 0.15$, $g_2 = 0.05$, $\sigma_1 = 0.6$, $\sigma_2 = 0.02$, $x = 0.01$, $x^\diamond = 1.5$, $\phi\left(x^\diamond\right) = 10$, and $\mu_0 = 0.01$

This section reconsiders the r/K selection theory by comparing various demographic indicators of optimal life history generated by adaptive strategies. Theorem 1 shows that the K-strategy has the largest density effect. In this model, the density effect is homogeneous with respect to age and size (Cf. Eq. (3.31)), which is equivalent to maximizing the carrying capacity (see Fig. 4.2). Thus, it is a special case that the K-strategy coincides with the strategy that maximizes the carrying capacity.

Let us compare the r-strategy with the K-strategy in the intrinsic rate of natural increase $\tilde{r}_0(\Gamma)$ in primary succession, which implies a sufficiently small density effect. As $\tilde{u}$ is constant, the $\tilde{r}_0(\Gamma)$ is obtained using the same procedure as that used to derive Eq. (2.58).

$$\tilde{r}_0(\Gamma) = \begin{cases} \rho_0\left[\theta_\kappa^1(\Gamma) - \frac{\sigma_1^2}{2}(1-\rho_0)\right] - \mu_0 & \tilde{u} = 0 \\ \rho_0\left[\theta_\kappa^1(\Gamma)(1-\tilde{u}) + \theta_{1-\kappa}^2(\Gamma)\tilde{u} \right. \\ \left. -\frac{1}{2}\left(\sigma_1^2(1-\tilde{u})^2 + \sigma_2^2\tilde{u}^2\right)(1-\rho_0)\right] - \mu_0 & 0 < \tilde{u} < 1 \\ \rho_0\left[\theta_{1-\kappa}^2(\Gamma) - \frac{\sigma_2^2}{2}(1-\rho_0)\right] - \mu_0 & \tilde{u} = 1 \end{cases}. \tag{4.49}$$

As Theorem 1 guarantees that the r-strategy maximizes r_0, let us compare it with r_0 in the K-strategy $\tilde{u}_K$ under a sufficiently small density $\Gamma \approx 0$.

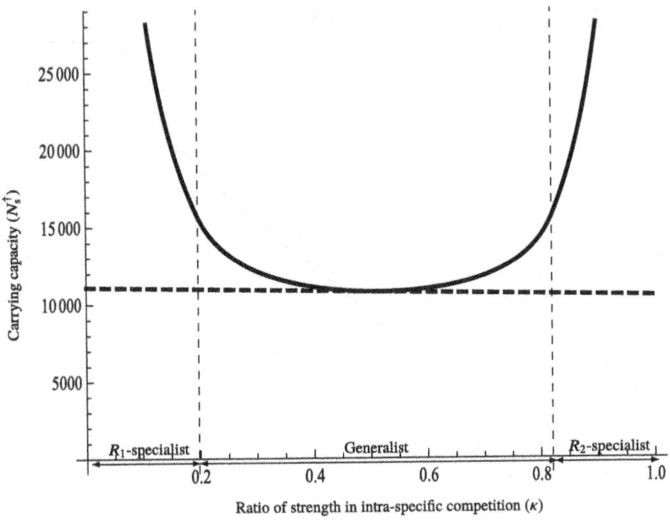

Fig. 4.2 Carrying capacities of r-strategy and K-strategy: the horizontal axis represents the ratio of the strength in the intra-specific competition (κ), and the vertical axis represents the size of the carrying capacity $\mathcal{N}^*$. The solid and dashed lines represent the K-strategy $\tilde{u}_{\mathrm{K}}$ and r-strategy $\tilde{u}_{\mathrm{r}}$, respectively. The parameters are the same as those used in Fig. 4.1. Each vertical dashed line represents the boundary between the specialists and the generalists. This figure shows that the K-strategy always has a greater carrying capacity than the r-strategy

$$\tilde{r}_0(\mathbf{o})\Big|_{\tilde{u}=\tilde{u}_{\mathrm{K}}} \leq \tilde{r}_0(\mathbf{o})\Big|_{\tilde{u}=\tilde{u}_{\mathrm{r}}}.$$

Figure 4.3 shows that the larger the bias in the density effect, the smaller is the intrinsic rate of natural increase of the K-strategy. In other words, it is thought that the K-strategy, i.e., the optimal strategy in the stationary population, rarely coincides with the r-strategy.

As shown in the analysis of the previous term, this specific model has constant adaptive control for the r-strategy and K-strategy. Constant adaptive control implies that we can compute several demographic indicators from the objective function derived by fixing the control u as a constant. Then, the objective function $v_r(x, \Gamma)$ is the solution of the following linear ODE:

$$\left[H^*(x, u, \Gamma) + r\right] v_r(x, u, \Gamma) + F_s(x) = 0.$$

$H^*(x, u, \Gamma)$ is given by Eq. (4.40).[4] As all the parameters except x are constant, we can solve this equation using the same method as that in Eqs. (2.57) and (4.45). The objective function $v_r(x, \Gamma)$ is equal to a change of parameters in Eq. (2.57) such that

[4] Note that we can treat u as a parameter in this case.

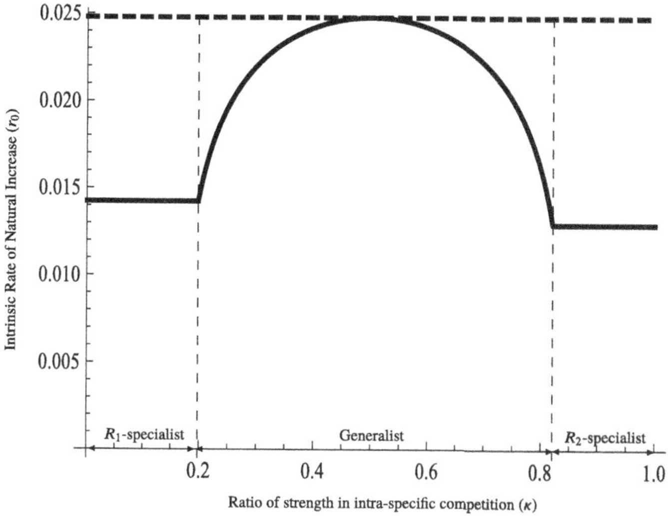

Fig. 4.3 Difference in the intrinsic growth rate of natural increase between r/K strategies: this figure illustrates the values of IRNI in the r-strategy and K-strategy. The horizontal axis denotes the ratio of strength in the intra-specific competition (κ), and the vertical axis denotes the IRNI. The solid and dashed lines represent the IRNI of the K-strategy and r-strategy, respectively. The curves are given by Eqs. (4.47)–(4.49). The IRNI of the r-strategy is always greater than that of the others based on its definition. The domains of both the specialists and the generalists represent the K-strategy because the r-strategy is independent of intra-specific competition. The parameters are the same as those used in Fig. 4.1

$$g_0 \to \theta_\kappa^1 (\Gamma) (1-u) - \theta_{1-\kappa}^2 (\Gamma) u, \quad \sigma_0 \to \sqrt{\left(\sigma_1^2 (1-u)^2 + \sigma_2^2 u^2\right)}.$$

Accordingly, $v_r (x, u, \Gamma)$ has the same functional form as Eq. (2.57):

$$v_r (x, u, \Gamma) = \phi \left(x^\diamond\right) \left(\frac{x}{x^\diamond}\right)^{\rho_+(r,u,\Gamma)} \tag{4.50}$$

Figure 4.4 shows the BRN for each strategy ($R_{0,r}$ and $R_{0,K}$ are BRNs for the r-strategy and K-strategy, respectively) under r-selection, which is given by

$$R_{0,r} = v_r (x, u, \Gamma)\Big|_{r=0, u=u_r, \Gamma=0}, \quad R_{0,K} = v_r (x, u, \Gamma)\Big|_{r=0, u=u_K, \Gamma=0}.$$

The K-strategy has the largest BRN, which is one, in a stationary population; however, it is not necessarily the largest in an environment with little or no density effects.

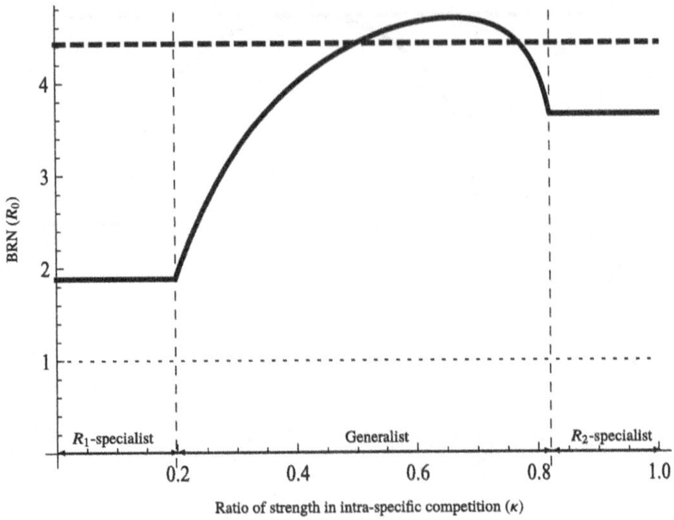

Fig. 4.4 Difference of BRN between r/K strategies under r-selection: this figure illustrates R_0 for both strategies in the absence of density effects. The horizontal axis represents the ratio of the strength in the intra-specific competition (κ), and the vertical axis represents R_0 under $\Gamma_t = 0$. The solid and dashed lines represent the same entities as those in Fig. 4.3, and the parameters used are also the same. It is shown that the K-strategy of R_0 is not always higher than the other. Therefore, the BRN of the K-strategy potentially depends on the ratio of the strength in the intra-specific competition

4.5.3 *Summary*

It is not reasonable to directly apply these results of the semelparous model to human demography. However, these results have many implications for human life history. In the absence of density effects, the individual stochasticity is controlled toward precocity and prolificacy. Meanwhile, in a saturated population, the individual stochasticity is controlled in the direction of preventing the invasion of other strategies. Then, the latter is possible for evolving the diversity of life histories because the control tendency depends on the ingredients of the life history affected by the density. Accordingly, the birth and death plans in human life history possibly influence the density of human society.

References

Taylor, H., Gourley, R., Lawrence, C., & Kaplan, R. (1974). Natural selection of life history attributes: an analytical approach. *Theoretical Population Biology, 5*(1), 104–122.

Leon, J. (1976). Life histories as adaptive strategies. *Journal of Theoretical Biology, 60*(2), 301–335.

Metz, J., & Diekmann, O. (1986). *The dynamics of physiologically structured populations* (Vol. 68). Springer.
Inaba, H. (2017). Age-structured population dynamics in demography and epidemiology.
Pianka, E. (1976). Natural selection of optimal reproductive tactics. *American Zoologist, 16*(4), 775–784.
Reznick, D., Bryant, M. J., & Bashey, F. (2002). r-and k-selection revisited: The role of population regulation in life-history evolution. *Ecology, 83*(6), 1509–1520.
Oizumi, R., Kuniya, T., & Enatsu, Y. (2016). Reconsideration of r/k selection theory using stochastic control theory and nonlinear structured population models. *PloS One, 11*(6), e0157715.
Øksendal, B. (2003). *Stochastic differential equations: An introduction with applications*. Springer.
Karatzas, I., & Shreve, S. (1991). *Brownian motion and stochastic calculus* (Vol. 113). Springer.
Iwasa, Y., & Roughgarden, J. (1984). Shoot/root balance of plants: Optimal growth of a system with many vegetative organs. *Theoretical Population Biology, 25*(1), 78–105.
Iwasa, Y., & Cohen, D. (1989). Optimal growth schedule of a perennial plant. *American Naturalist*, 480–505.
Oizumi, R., & Inaba, H. (2021). Evolution of heterogeneity under constant and variable environments. *PloS One, 16*(9), e0257377.
Bellman, R. (2013). *Dynamic programming*. Courier Corporation.
Yong, J., & Zhou, X. (1999). *Stochastic controls: Hamiltonian systems and HJB equations* (Vol. 43). Springer.
Stearns, S. (1992). *The evolution of life histories*. Oxford: Oxford Univeristy Press.
Aase, K. (1984). Optimum portfolio diversification in a general continuous-time model. *Stochastic Processes and their Applications, 18*(1), 81–98.
Shreve, S. (2004). *Stochastic calculus for finance: Continuous-time models* (Vol. 2). Springer.
Oizumi, R. (2014). Unification theory of optimal life histories and linear demographic models in internal stochasticity. *PloS One, 9*(6), e98746.

Chapter 5
Application to External Stochasticity

5.1 External Stochasticity and Perturbation Method

The previous chapters presented a parameter that maximizes the adaptive life history in a constant environment. This chapter presents the population dynamics behavior under a simple stochastic environment.

Although there are several assumptions and candidates for statistical noise as external stochasticity, we simplify the environmental stochasticity as white noise parameterized by a and y as $\mathscr{W}_t(a, y)$.

$$\mathscr{W}_t(a, y) = \frac{\partial}{\partial t}\mathcal{B}_t(a, y) \tag{5.1}$$

$$\mathbb{E}^{\text{ext}}[\mathscr{W}_t(a, y)] = 0 \tag{5.2}$$

$$\mathbb{E}^{\text{ext}}\left[\mathscr{W}_t(a, y)\mathscr{W}_t(a', y')\right] = \delta(a - a')\,\delta^d(y - y'), \tag{5.3}$$

for all $t > 0$. Here, $\mathcal{B}_t(a, y)$ denotes the Brownian motion parameterized by a and y. Assume that a population vector under external stochasticity $P_t^\epsilon(a, y)$ follows the stochastic PDE

$$\left(\frac{\partial}{\partial t} + \frac{\partial}{\partial a}\right) P_t^\epsilon(a, y) = -\left(H(a, y) - \epsilon\mathscr{W}_t(a, y)\right) P_t^\epsilon(a, y), \tag{5.4}$$
$$P_0^\epsilon(a, y) = p(a, y)$$

where ϵ denotes a sufficiently small positive constant that represents the strength of external stochasticity.

Let us consider the *long-term logarithmic growth rate* (LLGR) of the population denoted by $\bar{r}$ that is defined as

$$\bar{r} := \lim_{t\to\infty}\frac{1}{t}\ln\{\|P_t(\cdot)\|\}, \tag{5.5}$$

R. Oizumi, *Population Dynamics Based on Individual Stochasticity*,
Population Studies of Japan,
https://doi.org/10.1007/978-981-19-3548-0_5

where L^1-norm $\|f\|$ is defined as

$$\|f\| = \int_0^{\alpha} da \int_A dy\, f(a, y)\,.$$

From the renewal theorem given by Eq. (3.7), we have $\bar{r} = r_0$ in a constant environment.

Because it is difficult to compute a strict value of an LLGR involving external stochasticity, we apply a perturbation method to ϵ to calculate its approximate value, such that

$$\bar{r}(\epsilon) = r_0 + \epsilon D_1 + \epsilon^2 D_2 + \cdots\,.$$

Second-order Approximation of Long-term Logarithmic Growth Rate

We introduced the derivation of the second-order approximation of LLGR in Eq. (5.4). The population Hamiltonian vector, Hamiltonian, and noise functions are simplified to reduce the computational complexity as

$$\begin{aligned} P_t^{\epsilon} &= P_t^{\epsilon}(a, y) \\ P_0^{\epsilon} &= p(a, y) = p \\ \mathcal{W}_t &= \mathcal{W}_t(a, y)\,. \end{aligned}$$

Let us consider the following variation of the constant formula:

$$P_t^{\epsilon} = T(t)\,p + \epsilon \int_0^t dt'\, T\left(t - t'\right) \mathcal{W}_t' P_{t'}^{\epsilon}. \tag{5.6}$$

From Eqs. (2.38) and (2.40), the semigroup $T(t)$ is defined by

$$T(t)\,\varphi := \sum_{k=0}^{n} \langle v_k, \varphi\rangle \exp\{r_k t\}\, w_k + O(\exp\{(\Re r_k - \eta_0)t\}), \quad \eta_0 \geq 0.$$

With Eq. (5.6) and Ito's formula for multiple stochastic integrals (Itô, 1951), a perturbation of the population vector is found by iterative computation.

$$\begin{aligned} &P_t^{\epsilon} = \sum_{m=0}^{\infty} \varepsilon^m Q_m(t)\, p \\ &Q_0(t)\, p := T(t)\, p \\ &\int_A dx\, v_k(0, x)\, \nu(x) = \langle v_k, w_k\rangle^{-1} \\ &Q_m(t)\, p := \\ &\int_0^t \cdots \int_0^{t^{(m-1)}} dt' dt'' \cdots dt^{(m)}\, T\left(t - t'\right) \mathcal{W}_{t'} T\left(t' - t''\right) \mathcal{W}_{t''} T\left(t'' - t'''\right) \cdots \mathcal{W}_{t^{(m)}} T\left(t^{(m)}\right) p \end{aligned} \tag{5.7}$$

Using the symbol in Eq. (5.5), each $Q_m(t)\,p$ is deformed as

$$
\begin{aligned}
Q_m(t)\,p &= \sum_{k_1,\cdots,k_m}^{n} \exp\{r_{k_m} t\}\, w_{k_m} \\
&\quad \times \int_0^t \cdots \int_0^{t^{(m-1)}} dt' dt'' \cdots dt^{(m)} \prod_{\ell=1}^{m-1} \| v_{k_{m-\ell}} \mathscr{W}_{t^{(m-\ell)}} w_{k_{m-\ell+1}} \| \| v_{k_m} \mathscr{W}_{t^m} p \| \\
&= \sum_{k_1,\cdots,k_m}^{n} \exp\{r_{k_m} t\}\, w_{k_m} \int_0^t \cdots \int_0^{t^{(m-1)}} \prod_{\ell=1}^{m-1} \| v_{k_{m-\ell}} d\mathcal{B}_{t^{(m-\ell)}} w_{k_{m-\ell+1}} \| \| v_{k_m} d\mathcal{B}_{t^m} p \| \\
&= \frac{t^{\frac{m}{2}}}{m!} \sum_{k_1,\cdots,k_m}^{n} \exp\{r_{k_m} t\}\, w_{k_m} \| h_m\left(\frac{\mathcal{B}_t}{\sqrt{t}}\right) v_{k_m} p \prod_{\ell=1}^{m-1} v_{k_\ell} w_{k_\ell} \|,
\end{aligned}
\tag{5.8}
$$

where $h_m(x)$ denotes the Hermite polynomial

$$
h_m(x) := (-1)^m \exp\left\{\frac{x^2}{2}\right\} \frac{d^m}{dx^m} \exp\left\{-\frac{x^2}{2}\right\} \quad m = 0, 1, 2, \cdots.
$$

The last row in Eq. (5.8) is derived from the following formula (Itô, 1951):

$$
m! \int_0^t \int_0^{t'} \cdots \int_0^{t^{(m-1)}} dB_{t'} dB_{t''} \cdots dB_{t^{(m)}} = t^{\frac{m}{2}} h_m\left(\frac{B_t}{\sqrt{t}}\right) \quad t \ge t' \ge \cdots \ge t^{(m)}.
$$

The arbitrary constant of the adjoint eigenfunction is set as

$$
\int_A dx\, v_k(0, x)\, v(x) = \langle v_k, w_k \rangle^{-1} \quad k = 0, 1, 2, \cdots.
$$

If the population vectors in the presence and absence of external stochasticity are close to each other, $P_t^\epsilon \approx P_t^0 = P_t \quad (\epsilon \ll 1)$, the perturbation expressed in Eq. (5.8) provides an accurate approximation. With this assumption, an ϵ-specific mean LLGR $\bar{r}_{\mathrm{E}}(\epsilon)$ is represented by substituting Eqs. (5.8) into Eq. (5.5) such that

$$
\bar{r}_{\mathrm{E}}(\epsilon) = \lim_{t\uparrow\infty} \frac{1}{t} \mathbb{E}^{\mathrm{ext}}\left[\ln\{\|P_t^\epsilon\|\}\right] = \lim_{t\uparrow\infty} \frac{1}{t} \mathbb{E}^{\mathrm{ext}}\left[\ln\left\{\sum_{m=0}^{\infty} \epsilon^m \langle v_0, Q_m(t)\, w_0\rangle\right\}\right].
\tag{5.9}
$$

For simplicity, suppose that the initial population is the eigenfunction corresponding to the 0-th characteristic root.

By expanding Eq. (5.9) into a Taylor series at $\epsilon = 0$, the growth rate becomes

$$\bar{r}_{\mathrm{E}}(\epsilon) = r_0 + \lim_{t \uparrow \infty} \frac{\epsilon}{t} \mathbb{E}^{\mathrm{ext}} \left[\frac{\langle v_0, Q_1(t) w_0 \rangle}{\langle v_0, Q_0(t) w_0 \rangle} \right]$$
$$+ \lim_{t \uparrow \infty} \frac{\epsilon^2}{t} \mathbb{E}^{\mathrm{ext}} \left[\frac{\langle v_0, Q_2(t) w_0 \rangle}{\langle v_0, Q_0(t) w_0 \rangle} - \frac{1}{2} \left(\frac{\langle v_0, Q_1(t) w_0 \rangle}{\langle v_0^*, Q_0(t) w_0 \rangle} \right)^2 \right]$$
$$+ O\left(\epsilon^3\right). \tag{5.10}$$

Let us consider the mean growth rate in an environment consisting of sufficiently small disturbances such that the third-order (or higher-order) terms in ϵ can be truncated. The second term on the right-hand side is zero in the mean growth rate because of the statistical property of the fluctuation term (cf. Eq. (5.2)). Accordingly, the key point is the estimation of the second-order term in Eq. (5.10). One of the pieces constituting the second-order term is computed as

$$\langle v_0, Q_2(t) w_0 \rangle = \exp\{r_0 t\} \frac{t}{2!} \sum_k \| h_2 \left(\frac{\mathcal{B}_t}{\sqrt{t}} \right) v_0 w_0 v_k w_k \|. \tag{5.11}$$

The Hermite polynomials are orthogonal with respect to the Gaussian measure

$$\frac{1}{\sqrt{2\pi}} \int_{-\infty}^{\infty} dx\, h_m(x)\, h_{m'}(x) \exp\left\{ -\frac{x^2}{2} \right\} = m! \delta_{mm'},$$

i.e., the term becomes statistically zero.

$$\mathbb{E}^{\mathrm{ext}} \left[\langle v_0, Q_2(t) w_0 \rangle \right] = 0. \tag{5.12}$$

Similarly, the other component of the third term in Eq. (5.10) is computed as

$$\frac{1}{2} \lim_{t \uparrow \infty} \frac{1}{t} \mathbb{E}^{\mathrm{ext}} \left[\left(\frac{\langle v_0, Q_1(t) w_0 \rangle}{\langle v_0, Q_0(t) w_0 \rangle} \right)^2 \right] = \frac{1}{2} \lim_{t \uparrow \infty} \frac{1}{t} \int_0^t dt' \, \| (v_0 w_0)^2 \|$$
$$= \frac{1}{2} \| (v_0 w_0)^2 \|. \tag{5.13}$$

After combining the components (Eqs. (5.12) and (5.13)), the second-order approximation of the LLGR becomes

$$\bar{r}_{\mathrm{E}}(\epsilon) \approx r_0 - \underbrace{\frac{\epsilon^2}{2} \| (v_0 w_0)^2 \|}_{\text{deviation term}}. \tag{5.14}$$

This approximation is similar to the Tuljapurkar approximation (Tuljapurkar, 1982); however, it differs in several aspects. For instance, the deviation term corresponding to the original Tuljapurkar approximation is described by a sensitivity matrix. In this continuous version, the statistics concerning the diffusion process X_a account for the term. One important aspect is that the second term on the right-hand side of the equation above incorporates eigenfunctions. As described previously, the adjoint eigenfunction serves as an objective function to determine the adaptive strategy. This characteristic suggests that an adaptive species in a variable environment does not always maximize identical functions in a constant environment. Hence, we may find another adaptive strategy u^* as

$$\bar{r}_{\mathrm{E}}\left[\tilde{u}_r\right] \approx \tilde{r} - \frac{\epsilon^2}{2} \| \left(\tilde{v}_0 \tilde{w}_0\right)^2 \| \leq \bar{r}_{\mathrm{E}}\left[u^*\right],$$

where the arbitrary constant is set to

$$\int_A d\xi \; \tilde{v}_0\left(0, \xi\right) \nu\left(\xi\right) = \langle \tilde{v}_0, \tilde{w}_0 \rangle^{-1}.$$

Specific Model for Two-fold Stochasticity

The previous section of this chapter showed that the effect of the external stochasticity on the population growth is represented by the eigenfunctions corresponding to the dominant characteristic root in the mean environment. We use a specific mathematical model that is analytically solvable to examine the contribution of the internal stochasticity to the external stochasticity.

Let us consider the role of individual stochasticity in the external stochasticity. We construct a mathematical model that compares the LLGRs of a group of inhomogeneous growth rates with those of a group of homogeneous growth rates in a variable environment.

When $X_a \in \mathbb{R}_+$ is the size at age $a \in [0, \infty)$, as an effect of individual stochasticity, we reuse Eq. (2.46) as an example of this application. Heterogeneity is generated by the fluctuation in the second term on the right-hand side of Eq. (2.46). Mortality is constant. Then,

$$\mu\left(a, y\right) \equiv \mu_0 = const. \tag{5.15}$$

Fertility is assumed to be an allometric function in size,

$$F\left(a, y\right) = f_0 y^{\rho} \quad 0 < \rho < 1. \tag{5.16}$$

This life history generates the following Hamiltonian and adjoint Hamiltonian

$$H_{\mathrm{s}}\left(y\right) \varphi\left(y\right) = \mu_0 \varphi\left(y\right) - g_0 \frac{d}{dy}\left(y \varphi\left(y\right)\right) - \frac{\sigma_0^2}{2}\left(\frac{d^2}{dy^2} y^2 \varphi\left(y\right)\right)$$

in Eq. (2.53), respectively. Assuming that all newborns have an identical state x

$$\nu\left(y\right) = \delta\left(x - y\right),$$

the eigenfunction w_r satisfies

$$\left(\frac{\partial}{\partial a} + H_s(y) + r\right) w_r(a, y) = 0, \quad w_r(0, y) = \delta(x - y). \tag{5.17}$$

Substituting an ansatz

$$w_r(a, y) = \exp\{-(\mu_0 + r) a\} \phi(a, y)$$

into Eq. (5.17), the equation is converted into a Fokker–Planck equation,

$$\left(\frac{\partial}{\partial a} + H_s(y) - \mu_0\right) \phi(a, y) = 0, \quad \lim_{a \downarrow 0} \phi(a, y) = \delta(x - y),$$

which gives the PDF of the geometric Brownian motion in Eq. (2.46). The PDF is then given by the logarithmic normal distribution

$$\phi(a, y) = \frac{1}{y\sqrt{2\pi\sigma_0^2 a}} \exp\left\{-\frac{\left(\ln\frac{y}{x} - \left(g_0 - \frac{\sigma_0^2}{2}\right) a\right)^2}{2\sigma_0^2 a}\right\}.$$

Therefore, the eigenfunction is

$$w_r(a, y) = \frac{1}{y\sqrt{2\pi\sigma_1^2 a}} \exp\left\{-\frac{\left(\ln\frac{y}{x} - \left(g_0 - \frac{\sigma_0^2}{2}\right) a\right)^2}{2\sigma_1^2 a} - (\mu_0 + r) a\right\}. \tag{5.18}$$

Because the size growth rate follows an age-homogeneous Markovian process (Eq. (2.46)), this adjoint function does not depend on age.

$$\begin{aligned} v_r(a, y) =& v_r(0, x) \lim_{\alpha \uparrow \infty} \mathbb{E}_y \left[\int_a^\alpha ds\ f_0 X_s^\rho \exp\{-(r + \mu_0)(s - a)\}\right] \\ &= v_r(0, x) \lim_{\alpha \uparrow \infty} \int_a^\alpha ds\ \exp\{-(r + \mu_0)(s - a)\} \mathbb{E}_y\left[f_0 X_{s-a}^\rho\right] \\ &= v_r(0, x) \lim_{\alpha \uparrow \infty} \int_0^{\alpha - a} ds'\ \exp\left\{-(r + \mu_0) s'\right\} \mathbb{E}_y\left[f_0 X_{s'}^\rho\right] \\ &= v_r(x) \int_0^\infty ds'\ \exp\left\{-(r + \mu_0) s'\right\} \mathbb{E}_y\left[f_0 X_{s'}^\rho\right] \\ &= v_r(y). \end{aligned} \tag{5.19}$$

Therefore, we have

$$v_r(a, y) = v_r(y). \tag{5.20}$$

Then, the adjoint eigenfunction follows the adjoint equation

$$-\left(H_s^*(y)+r\right)v_r(y)+v_r(x)\,f_0y^{\rho}=0. \tag{5.21}$$

This equation is explicitly solvable using the following ansatz:

$$v_r(y)=C_ry^{\rho}\quad C\neq 0,$$

which gives

$$v_r(y)=C_ry^{\rho}=\frac{v_r(x)\,f_0y^{\rho}}{-g_0\rho+\frac{1}{2}\sigma_0^2\rho(1-\rho)+\mu_0+r}. \tag{5.22}$$

Because the function above can constitute the characteristic equation

$$\psi_r\Big|_{r=r_0}=v_r(x)\Big|_{r=r_0}=1, \tag{5.23}$$

the dominant characteristic root is computed as

$$r_0=\rho\left(g_0-\frac{1}{2}\sigma_0^2(1-\rho)\right)+f_0x^{\rho}-\mu_0. \tag{5.24}$$

By the definition of $0<\rho<1$, the characteristic root indicates that the internal stochasticity has a negative effect on the population growth in a constant environment

$$\frac{\partial r_0}{\partial\sigma}<0.$$

Hence, it is nonadaptive for species to have heterogeneity under the condition $0<\rho<1$.

Substituting the dominant characteristic root Eq. (5.24) into Eq. (5.22), the functional becomes

$$\langle v_0,w_0\rangle=v_r(x)\frac{d}{dr}\psi_r\Big|_{r=r_0}=\frac{1}{f_0x^{\rho}}.$$

Hence, the arbitrary constant is determined to be

$$v_0(x)=\langle v_0,w_0\rangle^{-1}=f_0x^{\rho},$$

and the adjoint eigenfunction corresponding to the dominant characteristic root is

$$v_0(y)=f_0y^{\rho}. \tag{5.25}$$

In this case, the adjoint eigenfunction corresponding to the dominant root matches the fertility function.

Combining the previous steps, the deviation term is given by

$$\frac{\epsilon^2}{2}\| (v_0 w_0)^2 \| = \frac{\epsilon^2}{2}\int_0^\infty \int_{\mathbb{R}_+} dady \, \frac{f_0^2 y^{-2(1-\rho)}}{2\pi\sigma_0^2 a} \times \exp\left\{ -\frac{2\left(\ln\frac{y}{x} - \left(g_0 - \frac{\sigma_0^2}{2}\right)a\right)^2}{2\sigma_0^2 a} - 2(\mu_0 + r)a \right\}\Bigg|_{r=r_0}$$
$$= \frac{\epsilon^2 f_0^2 x^{2\rho-1}}{4\sigma_0\sqrt{g_0\left(\rho\left(2-\sqrt{2}\right)+\frac{\sqrt{2}}{2}\right)+\frac{1}{2}\sigma_1^2\left(\sqrt{2}\rho+\frac{\sqrt{2}-1}{2}\right)+f_0x^\rho}}. \tag{5.26}$$

This deviation term diverges to infinity in the absence of internal stochasticity:

$$\sigma_1 \to 0, \quad \frac{\epsilon^2}{2}\| (v_0 w_0)^2 \| \to \infty.$$

Thus, it is reasonable to consider the effect of higher orders of ε on this divergence. However, this consequence suggests the significance of heterogeneity in the persistence of species in minimally variable environments. This property contradicts the effect of internal stochasticity on the dominant characteristic root (Eq. (5.24)). Substituting Eqs. (5.24) and (5.26) into Eq. (5.14), the LLGR approximates

$$\bar{r}_E(\epsilon) \approx \rho\left(g_0 - \frac{1}{2}\sigma_1^2(1-\rho)\right) + f_0 x^\rho - \mu_0 - \frac{\epsilon^2 f_0^2 x^{2\rho-1}}{4\sigma_0\sqrt{g_0\left(\rho\left(2-\sqrt{2}\right)+\frac{\sqrt{2}}{2}\right)+\frac{1}{2}\sigma_0^2\left(\sqrt{2}\rho+\frac{\sqrt{2}-1}{2}\right)+f_0x^\rho}}. \tag{5.27}$$

The LLGR represents a monotonically increasing function with respect to the mean size growth rate g_0,

$$\frac{\partial \bar{r}_E(\varepsilon)}{\partial g_0} \geq 0.$$

This point may appear to be trivial; however, it is notable that the deviation term monotonically decreases in g_0. Further, rapid growth may reduce the risks inherent in variable environments.

The heterogeneity of the size growth rate reduces the mean dominant characteristic root and causes the risk of extinction from the variable environment. By computing Eq. (5.27) in terms of σ and ϵ, we find the adaptive heterogeneity of the size growth rate for each ϵ (see Fig. 5.1).

Figure 5.1 shows that adaptive heterogeneity increases with environmental variability. The numerical analysis suggests that species evolve to yield heterogeneity in

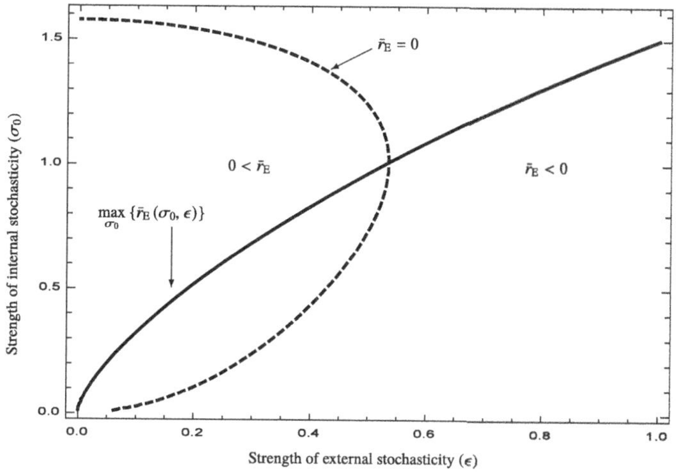

Fig. 5.1 Adaptive heterogeneity under two-fold stochasticity: This figure shows the existence of an adaptive value in σ_0. Each ϵ representing the strength of external stochasticity has a unique adaptive value of σ_0 that maximizes the dominant characteristic root r_0; ϵ increases with an adaptive value σ_0. This result suggests that species require greater heterogeneity in more variable environments. The parameters are $g_0 = 0.6$, $x = 0.01$, $\mu_0 = 0.1$, $f_0 = 1.0$, and $\rho = 0.4$

variable environments. This viewpoint corroborates conventional interpretations of the necessity for biodiversity.

Adaptive Resource Utilization in External Stochasticity

Based on Eqs. (1.4), (5.15), and (5.16), we consider the adaptive resource utilization in Eq. (4.39) under external stochasticity. For simplicity, we deal with the r-strategy ($\Gamma = 0$). Then, the HJB equation becomes

$$\inf_{u \in [0,1]} \left\{ \left[H^* (x, u) + r \right] \tilde{v}_r (x) \right\} + f_0 x^\rho = 0, \tag{5.28}$$

where the Hamiltonian $H^* (x, u)$ is the same as Eq. (4.40). Solving Eq. (5.28) using the same procedure as that used in the derivation of Eq. (4.48), we get

$$\tilde{v}_r (x) = \frac{f_0 x^\rho}{r - \lambda}$$

$$\lambda = \begin{cases} \rho \left(g_1 - \frac{1}{2}\sigma_1^2 (1-\rho) \right) + f_0 x^\rho - \mu_0 & \tilde{u}_r = 0 \\ \left(\frac{g_1 \sigma_2^2 + g_2 \sigma_1^2}{\sigma_1^2 + \sigma_2^2} \right) \rho - \frac{1}{2} \frac{\sigma_1^2 \sigma_2^2 \rho (1-\rho)}{\sigma_1^2 + \sigma_2^2} + \frac{1}{2} \frac{(g_1 - g_2)^2 \rho}{\left(\sigma_1^2 + \sigma_2^2\right)(1-\rho)} + f_0 x^\rho - \mu_0 & 0 < \tilde{u}_r < 1 \\ \rho \left(g_2 - \frac{1}{2}\sigma_2^2 (1-\rho) \right) + f_0 x^\rho - \mu_0 & \tilde{u}_r = 1 \end{cases},$$

and the adaptive utilization is

$$\tilde{u} = \max\left\{\frac{\sigma_1^2}{\sigma_1^2+\sigma_2^2} - \frac{(g_1-g_2)}{(\sigma_1^2+\sigma_2^2)(1-\rho)}, 0\right\}. \tag{5.29}$$

This control indicates that R_2-specific utilizers do not evolve. Because adaptive utilization is constant in constant environments, finding another utilization constant u^* that maximizes the LLGR in a variable environment implies that another adaptive utilization exists, even if the constant does not imply optimal control. Suppose that the utilization is always constant and that u^* becomes the adaptive strategy for two-fold stochasticity. The utilization constant specific LLGR $\bar{r}[u](\varepsilon)$ considers whether the variable environment selects a life history that favors heterogeneity as adaptive. Because the utilization rate does not depend on age or size, we can consider a specific LLGR with the following change of coefficients in Eq. (5.27).

$$g_0 \to g_1(1-u) + g_2 u, \quad \sigma_0 \to \sqrt{\sigma_1^2(1-u)^2 + \sigma_2^2 u^2}.$$

Solving $\bar{r}[u](\epsilon)$ numerically, we find that a variable environment favors heterogeneity, as suggested in the previous section of this chapter (Fig. 5.1). This figure shows adaptive resource utilization with respect to ρ for several values of ϵ. Although ρ represents the scaling exponent of fertility that denotes a measure of risk aversion in a deterministic environment, the adaptive utilization of risk appetite in the presence of external stochasticity exists in the domains of greater and smaller values of ρ (see Fig. 5.2). The domain of the adaptive strategy utilizing both resources becomes

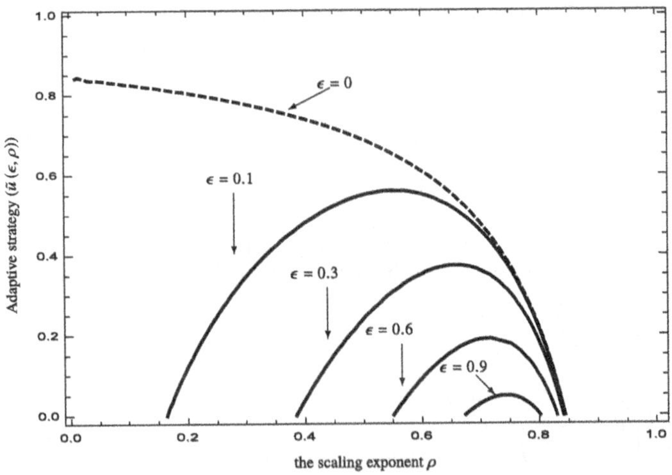

Fig. 5.2 Adaptive resource utilization under external stochasticity. The solid lines represent u^* for each value of ϵ, and the dashed lines represent the adaptive control given by Eq. (5.29) ($\epsilon = 0$). The parameters are $g_2 = 0.6$, $g_2 = 0.5$, $\sigma_1 = 0.8$, $\sigma_2 = 0.005$, and $\epsilon = \{0, 0.1, 0.3, 0.6, 0.9\}$; the others are the same as those in Fig. 5.1

narrower as ϵ increases. This consequence is linked to the relationship between the adaptive value of σ_0 and ϵ in Fig. 5.1.

The scaling exponent ρ represents the risk appetite index in economics; small values favor risk aversion. The adaptive strategy given by Eq. (5.29) represents an interpretation of the exponent that is identical to that in economics. However, under external stochasticity $\epsilon \neq 0$, minimal internal stochasticity does not become adaptive for small values of ρ.

References

Itô, K. (1951). Multiple wiener integral. *Journal of the Mathematical Society of Japan, 3*(1), 157–169.

Tuljapurkar, S. (1982). Population dynamics in variable environments. iii. evolutionary dynamics of r-selection. *Theoretical Population Biology, 21*(1), 141–165.

Epilogue: Population Dynamics from the Perspective of Individual Stochasticity

There is much stochasticity in the environment surrounding us, including external stochasticity, such as variable environments that affect the entire population, and internal stochasticity caused by the life histories of individuals. In this book, we focused on the latter and investigated the mathematics of life history and population dynamics. Individual stochasticity affects the population growth rate and total stationary population.

In a life history model based on SDEs, the impact of individual probabilities on the population dynamics varies with the type of stochastic integration employed; the results differ depending on whether Ito's or Stratonovich's integral is used. This book employed Ito's integral by considering models using Markov processes in demography and ecology.

The age–multi-state distribution using the path integral representation (see Eqs. (2.15) and (2.16)) explicitly expresses the effect of individual stochasticity. The path that minimizes the Lagrangian becomes the most observable growth curve. Therefore, each growth curve follows the statistical properties given by the SDE; moreover, it is advantageous for survival to pass through a state with a lower mortality rate. In a population with individual probability, the observed growth curve of an individual is affected by the mortality rate. This phenomenon never occurs in a deterministic model having a unique solution for the initial condition.

Life histories involving individual stochasticity can be incorporated into an age–multi-state structured model. In the absence of density effects, the model behaves in a Malthusian asymptotic manner. The characteristic equation gives the rate of increase. We analyzed the effect of individual stochasticity on the population dynamics via the characteristic functions that constitute this characteristic equation. In particular, when discussing evolution, the optimal life history can be derived by solving the control equation (HJB equation) derived from the equation of the reproductive value. The central argument is that the optimal life history maximizes the function that gives the reproductive value, regardless of the presence of density effects. The evolution of individual stochasticity also plays an essential role in this maximization.

R. Oizumi, *Population Dynamics Based on Individual Stochasticity*,
Population Studies of Japan,
https://doi.org/10.1007/978-981-19-3548-0

In addition to general formulations, examples of models that can be solved analytically were presented to make the book enjoyable for readers. Although these examples are not necessarily models of natural organisms, they provide many insights. For example, the semelparity model shows the effect of individual stochasticity on the population dynamics. This model indicates the conditions under which individual stochasticity negatively affects the population growth rate. Furthermore, intraspecific competition was shown to reduce the carrying capacity. In this model, the optimal mature size is determined only by the initial size and the number of births at maturity. In addition, the stability of the stationary population can be shown analytically. These properties simplify the impact of individual stochasticity on several demographic indicators.

The characteristics of the semelparity model also have implications for the optimal life schedule problem regarding individual stochasticity. In deterministic theory, faster size growth of individuals increases their reproductive success. Meanwhile, under individual stochasticity, the statistical size growth rate and its variance affect the reproductive success. Consequently, strategies can evolve to utilize nutrients and resources with low foraging risk. Such risk-averse strategies are also not deterministic. Furthermore, when density effects are added, the control becomes more diverse. This diversity is because the strategy with large density effects becomes dominant in a stationary population.

In a reconsideration of the r/K-selection theory, K-strategies can have more diverse values of demographic indicators than the conventional classification. This is because the density effect can affect mortality, reproduction, and life history in general. Nevertheless, evolution only maximizes the function that gives a reproductive value. Therefore, the K-strategy excludes other strategies by an uninvadable strategy on the density effect, to use the terminology of adaptive dynamics. The values of the demographic indices that differ from those of the r-strategy are considered as byproducts.

External stochasticity has been studied extensively in structured population models formulated by matrices. Tuljapurkar (1982) conducted a representative study on the argument that external stochasticity increases the risk of extinction (Tuljapurkar, 1982). Such studies are flourishing in ecology. The applications presented in this book are infinite-dimensional versions of Tuljapurkar's work. Moreover, the framework presented in this book can be used to handle the life history evolution that matrix models cannot represent. Individual stochasticity can also be shown to reduce the risk of extinction and population decline in variable environments, as shown in Figs. 5.1 and 5.2. Meanwhile, it must be noted that this model is still in its infancy; Eq. (5.4) cannot guarantee a monotonous non-increase in the cohort. This rule is because the number of cohorts must either be conserved or decrease with age in a blockade population with no population inflow.

However, this framework faces many challenges, such as life history evolution in variable environments, resource allocation across generations, and interactions with other species. The methods used in this book have not addressed these challenges thus far. The author has recently conducted some suggestive studies on the stochasticity of individuals in variable environments; however, they are a type of approximation and may not be sufficient. For example, in this book, the initial state of newborns is given independent of the fertility; however, for some animals and plants, including humans, it is natural that the state of the mothers may affect the initial state of the offspring. In this case, the reproductive value term, an arbitrary constant, becomes an integral equation and is difficult to solve because the initial value information is included in the birth rate. In addition, we may consider the bisexual problem or differential games in life history evolution for other species and their interactions. This also involves many themes and demographic conundrums.

In summary, although there are many unsolved problems even within the scope of the author's thinking, it would be better if the reader is interested in these problems and develops the contents of this book accordingly. However, even if readers are not interested, we would be motivated to interest them in this book.

Acknowledgement We are deeply indebted to Hisashi Inaba and Toru Suzuki for their insightful comments and advice on this book. The author thanks the Springer staff (Kavitha Palanisamy, Yutaka Hirachi, and Shinko Mimura) for their patience with the author's repeated delays in submitting the manuscript. Further, we thank Kensaku Kinjo, Kota Hattori, and Youichi Enatsu for their insightful comments and advice on this study, and Kumiko Oizumi, Shin Oizumi, Koh Oizumi, and Hiroko Oizumi for their support and encouragement. We would also like to thank Editage (www.editage.com) for English language editing. This research was supported by the Japan Society for the Promotion of Science (JSPS) KAKENHI (Grant Number 20K14368) and the Ministry of Health, Labour and Welfare (Grant number 20AA2007).

Reference

Tuljapurkar, S. (1982). Population dynamics in variable environments. iii. evolutionary dynamics of r-selection. *Theoretical Population Biology*, *21*(1), 141–165.